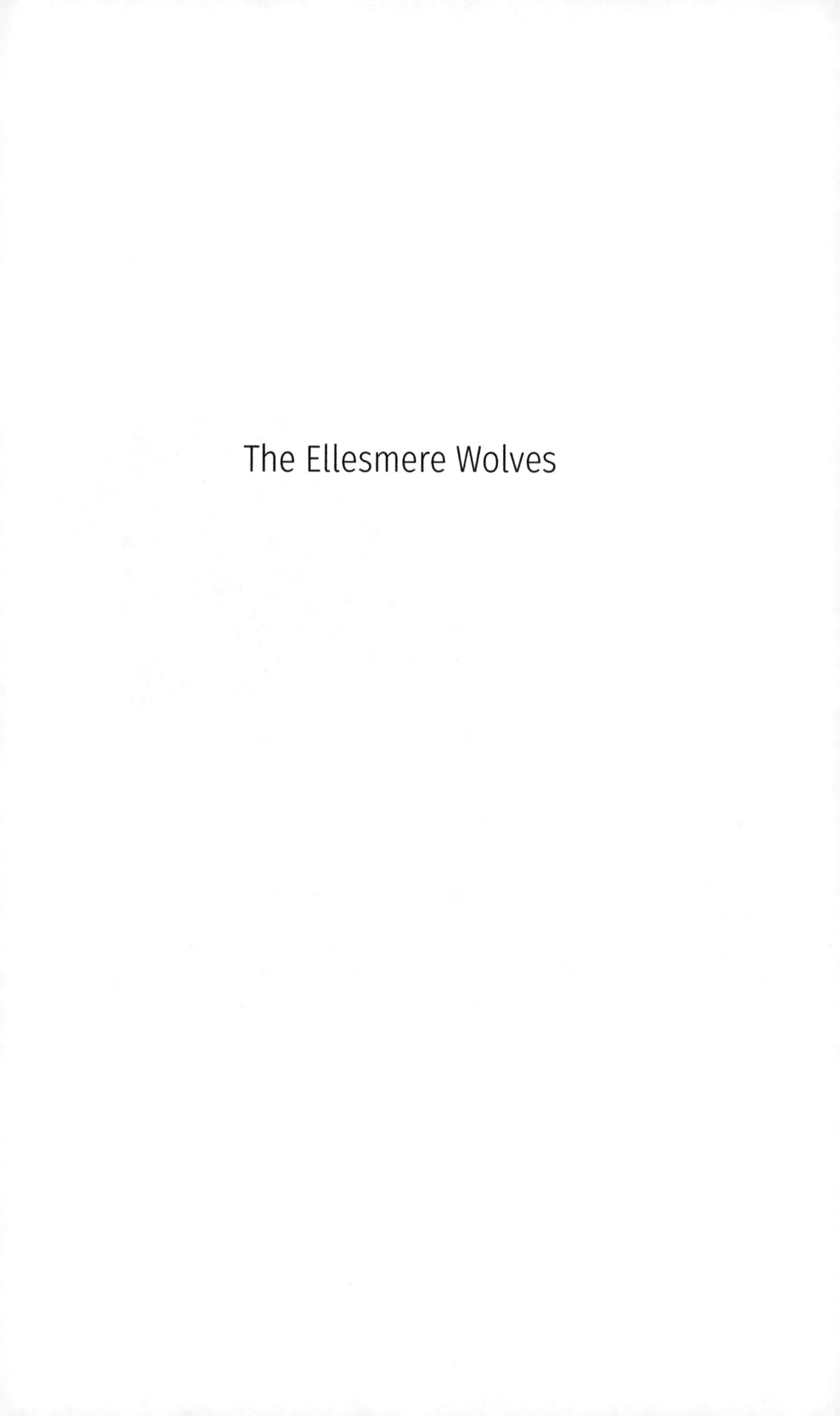

The Ellesmere Wolves

THE Ellesmere Wolves

Behavior and Ecology in the High Arctic

L. David Mech, Morgan Anderson,
and H. Dean Cluff

The University of Chicago Press
Chicago and London

The University of Chicago Press, Chicago 60637
The University of Chicago Press, Ltd., London

Published 2025
Printed and bound by CPI Group (UK) Ltd, Croydon, CR0 4YY

34 33 32 31 30 29 28 27 26 25 1 2 3 4 5

ISBN-13: 978-0-226-83372-9 (cloth)
ISBN-13: 978-0-226-83374-3 (paper)
ISBN-13: 978-0-226-83373-6 (e-book)

DOI: https://doi.org/10.7208/chicago/9780226833736.001.0001

Library of Congress Cataloging-in-Publication Data

Names: Mech, L. David, author. | Anderson, Morgan, author. | Cluff, H. Dean (Howard Dean), 1961– author.
Title: The Ellesmere wolves : behavior and ecology in the high Arctic / L. David Mech, Morgan Anderson, and H. Dean Cluff.
Description: Chicago : The University of Chicago Press, 2025. | Includes bibliographical references and index.
Identifiers: LCCN 2024029772 | ISBN 9780226833729 (cloth) | ISBN 9780226833743 (paperback) | ISBN 9780226833736 (e-book)
Subjects: LCSH: Wolves—Nunavut—Ellesmere Island. | Human-animal relationships.
Classification: LCC QL737.C22 M3994 2025 | DDC 599.77309719/52—dc23/eng/20240705
LC record available at https://lccn.loc.gov/2024029772

♾ This paper meets the requirements of ANSI/NISO Z39.48-1992 (Permanence of Paper).

WE DEDICATE THIS BOOK

to the following colleagues and other wolf workers whom we have known personally, who have already passed but who we believe would all have thoroughly enjoyed the experiences we describe in this book: Durward Allen, Glenn DelGiudice, Gordon Haber, Bob Hayes, Mark Korb, Tom Meier, Don Murray, John Parker, Roger Peters, Doug Pimlott, Erkki Pulliainen, Jeff Renneberg, Ulie Seal, Bob Stephenson, Roger Stradley, Vic Van Ballenberghe, and Bob Wayne.

Contents

Foreword LUIGI BOITANI *ix*
Preface *xiii*

1. Lifetime Highlight *1*
2. The Den *10*
3. Ellesmere Island *16*
4. Living with the Pack *21*
5. The Plot Thickens *31*
6. A Flourishing Family Falters *38*
7. A Whole New Phase *53*
8. Beyond Wolf Behavior *64*
9. South of Slidre *74*
10. Cast of Characters *84*
11. Ellesmere Dens *89*
12. It's All about Pups *98*
13. Many Mouths to Feed *109*

14. Food Caching *116*

15. Big Burly Beasts *124*

16. Hares, Caribou, and Seals *136*

17. The Daily Hunt *150*

18. Divisions of Labor *164*

19. Life at Wolf Headquarters *175*

20. Just for the Howl of It *184*

21. Territoriality and Scent-Marking *191*

22. The Bigger Picture *206*

23. Climate Change *209*

24. Forever Wild *214*

Acknowledgments 219
Appendix I 221
Appendix II 227
Literature Cited 229
Index of Names 247
Index of Subjects 251

Color photographs follow page 112.

Foreword

LUIGI BOITANI

This is a remarkable book. The wolf literature is already huge with several thousand scientific papers and a few dozen books, evidence of the wolves' tremendous ecological flexibility in adapting to almost all terrestrial types of environments, from the extreme arctic tundra to the scorching sands of the Arabian Peninsula. The flow of new scientific information continues today and appears to be endless, helped by new technologies and field methods that allow us to explore ever more sophisticated aspects of wolves' biology and ecology. Long-term research such as the Yellowstone Wolf Project has set new standards of data quantity and quality that will be difficult to challenge. In this exciting landscape of scientific products, this book sets a completely different standard of excellence and is truly remarkable for many reasons.

I think many zoologists working with wild animals would agree that one of our dreams is to be able to live with our studied animals during all phases of their life, when they sleep, hunt, play, mate, nurse their pups, and more. The dream is to be able to be close to our animals but invisible to them, being able to watch them continuously for hours and days without interfering with their behavior. Dave Mech and his colleagues had the extraordinary opportunity to live that dream without the need to be invisible. The unique situation of wolves habituated to human presence allowed the scientists to observe them, live with them, and follow them during all activities for days and days. New facts and insights suddenly emerge as obvious when you can watch animals in their natural context: how else can we truly understand, for example, the dynamics among pack members?

An important difference sets this study of the Ellesmere wolves apart from other studies on habituated animals such as many primate species: those wolves ignored the observers, although they were certainly aware of them and occasionally interacted with them. The wolves were not frightened of people, did not include the scientists in the web of their social units, and did not "accept" them. In short, there is no indication that human presence had any effect on the wolves' daily life, a privilege that any scientist working with wild animals would dream of, the nearest possible condition to being invisible. Moreover, Ellesmere's habitat is treeless and open, in a series of flat and gently rolling hills, so you can watch animals at great distances without distracting events. The pure white of the wolves is immediately evident, even from a distance, against the uniform gray of the tundra.

Another reason for the uniqueness of the book is the study area. We all know that wolves live from the High Arctic regions to the edge of tropical forests, but it needs a special effort to imagine what life can be for a wolf living only a few hundred miles from the North Pole. Winter temperatures often below –40°F (–40°C), frozen ground and darkness for several months, scarcity of prey, and a monotonous habitat are serious challenges for any living creature. Reading through the chapters of this book, you will find data shedding new light on many fundamental aspects of wolf biology, like, for example, the dynamics of pack composition, social hierarchy, predation, and more; but you'll also find much new information on how wolves cope with the extreme environment of their island. The wolf can be many different animals depending on where the wolves are observed, as their ecology and behavior change as they fit the local latitude, habitat type, prey species, hybridization with other canids (e.g., coyote, jackal, dog), and behavior in relation to human activities. Thus, the challenge for any researcher is in seeing the fundamentals of wolf biology through the local details and to keep them distinct, a challenge masterfully overcome by this book.

The last, and perhaps the most important, reason that sets this book apart in the crowded shelf of wolf books is its authors, and especially its first author. To get the most out of the opportunity to be "invisible" near your subject animals, you need to have all the best possible skills in observing animals, all possible experience in studying wolves in the wild, and all possible knowledge of wolf biology. None could use the opportunity better than Dave Mech, who has the widest understanding of wolf biology of any biologist on Earth, having worked on wolves more than any other scientist and in a variety of wolf populations and geographic ranges. The combined skills of the three authors have extended the research scope on Ellesmere through additional techniques (e.g., radiotelemetry, genetics) beyond the direct observation of wolves and have produced a uniquely robust description of the wolves of the High Arctic. The competent, meticulous, perseverant, and candid attitude of the authors has been instrumental in collecting, digesting, and working on a huge quantity of data and making sense of them in a consolidated view of what it means to be a wolf of the arctic. But I know that the fieldwork of all three authors has been more than an exciting scientific assignment: no matter the science that has been produced, it has been a journey through the deepest reasons for the love of and commitment to studying wolves. This was the impact of Ellesmere wolves on me when I joined Dave on the island for one summer season, and it changed my own life.

Reading through the pages of this book, you will find exciting stories unfolding from a long series of scrupulous description and quantification of wolves' activities, movements, and interactions, some of which may even be boring to those who are not used to the painstaking work of taking notes of every detail essential to compose the overall grand picture. The key story of the book is one of incredible adaptability, stamina, resilience, and endurance: the wolf is all this, well beyond the stereotyped simplifications of the wolves living in any particular region of the world. The Ellesmere wolves show that there is much more, in a wolf, than what we used to believe when we look at our country, whether North America or Europe or Asia.

So I invite you to read this remarkable book with a curiosity for the biological details (and you will find tons of them) and a fascination for the magnificent vision of a pack of white phantoms even in the dark of the arctic winter.

Preface

The Ellesmere wolves are special, really special.

"Ellesmere" refers to Ellesmere Island, a large mostly frozen land mass whose northern tip is just 475 miles (765 kilometers) from the North Pole. The wolves on this High Arctic Canadian landscape are special, not only because they represent the northernmost population of the species, but primarily because they are tame, that is, unafraid of people. Those wolves evolved in that inhospitable landscape for centuries without any humans around (Mech and Janssens 2022).

Contrary to just about everywhere else in the wolf's circumpolar distribution where these large carnivores competed with humans for the same prey, on much of Ellesmere they were unfamiliar with people. A few Thule folks hunted along some of the shores centuries ago but not over the vast inland, much of which was perpetual ice and snow. Even more recent human settlements like airbases and weather stations clung to the shoreline.

Only during the last century did explorers and researchers begin to traverse the inner reaches of Ellesmere. When they did, they were surprised to find wolves that paid them little attention, except maybe to approach in curiosity. Elsewhere in the world, most wolves fled from even the scent of humans so quickly that those folks never even saw the animals.

Not so the Ellesmere wolves, as well as those on other islands in the Queen Elizabeth Archipelago of the High Arctic. Those wolves were essentially tame when it came to interacting with people. The potential for studying such wolves was obvious, for researchers would be able to observe close-up the behavior of individual wolves interacting with their pack mates as well as hunting their prey.

It was inevitable that eventually scientists studying wolves would learn of the unique High Arctic wolf population and seek to study it. I was one such individual, a US government employee who had begun my wolf investigations on famed Isle Royale in Lake Superior some 65 years ago. While then immersing myself in a career of studying wolves and deer in the Superior National Forest of Minnesota, I was also developing and refining various large-carnivore study techniques.

Eventually I was called upon to teach these techniques to graduate students and colleagues in regions from Italy, Africa, and India to Canada and Alaska, not only for use on wolves but on leopards, lions, tigers, and other exotic species. I was not necessarily in need of any other adventure. However, I had learned about the white arctic wolves on Ellesmere Island while researching my 1970 book, *The Wolf: The Ecology and Behavior of an Endangered Species.*

Since then, my long professional and personal quest had been to visit that far-off area and observe these unusually approachable wolves. But Ellesmere is so distant and out of reach that one could not just get on a plane and head there even if funds were available. Besides the astronomical cost and difficult weather conditions, considerable bureaucratic red tape also hinders one's ability to get there. Not until a *National Geographic* contract to chronicle Ellesmere Island materialized in 1986 did I get the chance.

What I found after photographer Jim Brandenburg and I arrived at −35°F (−37°C) in the middle of the April 11 night—brightened by the season's perpetual sunlight—became a cover story for *National Geographic.* It also morphed into 24 consecutive summers of investigating the Ellesmere wolves up close and personal.

In addition, while finally winding down my long stint on Ellesmere, I turned it over to former student Dan MacNulty. Dan, with colleagues Morgan Anderson and Dean Cluff, spent parts of a few more summers there studying wolves via collaring them with GPS radio collars. This new technology allowed the team (and me virtually) to gain a broader view of the wolf population I had spent so much time with.

My colleagues were well suited for this assignment. Dan had studied hunting behavior in Yellowstone wolves for both his master's and PhD degrees, recently become a professor at Utah State University, and was wide open to new wolf research. Dean had been an arctic hand since 1984, dealing with caribou, moose, bison, muskoxen, wolverine, and black and grizzly bears as a Northwest Territories biologist. He was familiar with Ellesmere from his studies of belugas and polar bears and had been studying wolves since 1995 and joining me on Ellesmere several summers from 2004 through 2010.

Morgan was a well-experienced add-on, having studied wolves in six areas, including collaring some 50 or so. She had worked extensively with the Inuit communities of Resolute Bay and Grise Fiord on priority wildlife monitoring and management projects and became key to the team gaining official approval to GPS-collar Ellesmere wolves. Morgan and Dean gladly joined me in preparing this book, while Dan, with many graduate students to tend to, was unable to find the time to collaborate.

Following is my year-by-year scientific adventure interacting with wolves so intimately that one tried to untie my bootlace. Included for completeness are data and findings from my compatriots who followed up on my studies in this unique study area. With wolves recolonizing several of the states in the United States (Mech 2017b) as well as several countries in western Europe (Chapron et al. 2014) over the past several decades, the need for more information about wolves was clear.

:: 1 ::

Lifetime Highlight

> It was the highlight of my life. Hundreds of miles north of Hudson Bay, a thousand or more miles from the nearest city, I stood alone in the High Arctic—surrounded by wolves. No question. I had just discovered their den of pups, and they were only fifty feet away. All my searching had paid off. All the hard hours of hiking, all the planning, the hoping and dreaming. After twenty-eight years, I had finally scored the Big One.

So I had written in my 1988 book, *The Arctic Wolf: Living with the Pack* (and see fig. 1.1).

I was standing on Ellesmere Island in the Canadian territory now known as Nunavut, some 2,500 miles (4,000 kilometers) from my Minnesota home and less than a quarter of that from the North Pole (fig. 1.2). Now, with seven white arctic wolves barking and howling around me, disturbed that I was near their den, I was in my glory. I had found many wolf dens before, but they were all seen from an aircraft, and all were in forested areas where if one got close enough to observe the pups, that would be too close; the adults would move the pups away. I had longed to study the behavior and interactions of wolves and their pups like Adolph Murie had done and described in his 1944 book, *The Wolves of Mount McKinley*.

Sure, there was Murie's (1944, 29) note that "the routine activity at the den was unexciting and quiet. For 3 or 4 hours at a time there might not be a stir." I would later find that instead of 3 or 4 hours, I could substitute 8 to 10 hours. Still, Murie also followed his statement with the words, "Yet it was an

Figure 1.1. Now after a long week of trying to find their den of pups, I was met by a greeting committee of upset adults.

inexhaustible thrill to watch the wolves simply because they typify the wilderness so completely." I was now living that thrill.

Most of the 28 years of my career as a wolf biologist at that moment had been spent studying wolves in Minnesota, although I had done my PhD research in Isle Royale National Park, Michigan (Mech 1966), and spent stints in Italy, Manitoba, and Alaska teaching other wolf biologists various study techniques such as live-trapping and radio-tracking wolves. Although these techniques are standard now, they were pioneering then. I had also observed and studied wolves in a captive colony for many years.

Although I had learned much about wolf predation on various prey, and about wolf movements, territoriality, mortality, survival, and physiology, I had never been in a situation where I could watch wild wolves from the ground behaving naturally for any period. Even special trips had failed, including trips to the north slope of the Brooks Range in Alaska and to Bathurst Inlet in Canada's Northwest Territories to try to locate a wolf den above the tree line where I could observe wolf behavior through a spotting scope.

Now I was again above the tree line—way above it, like 1,550 miles (2,500 kilometers) above it (fig. 1.3). The den I had just discovered was a cave-like

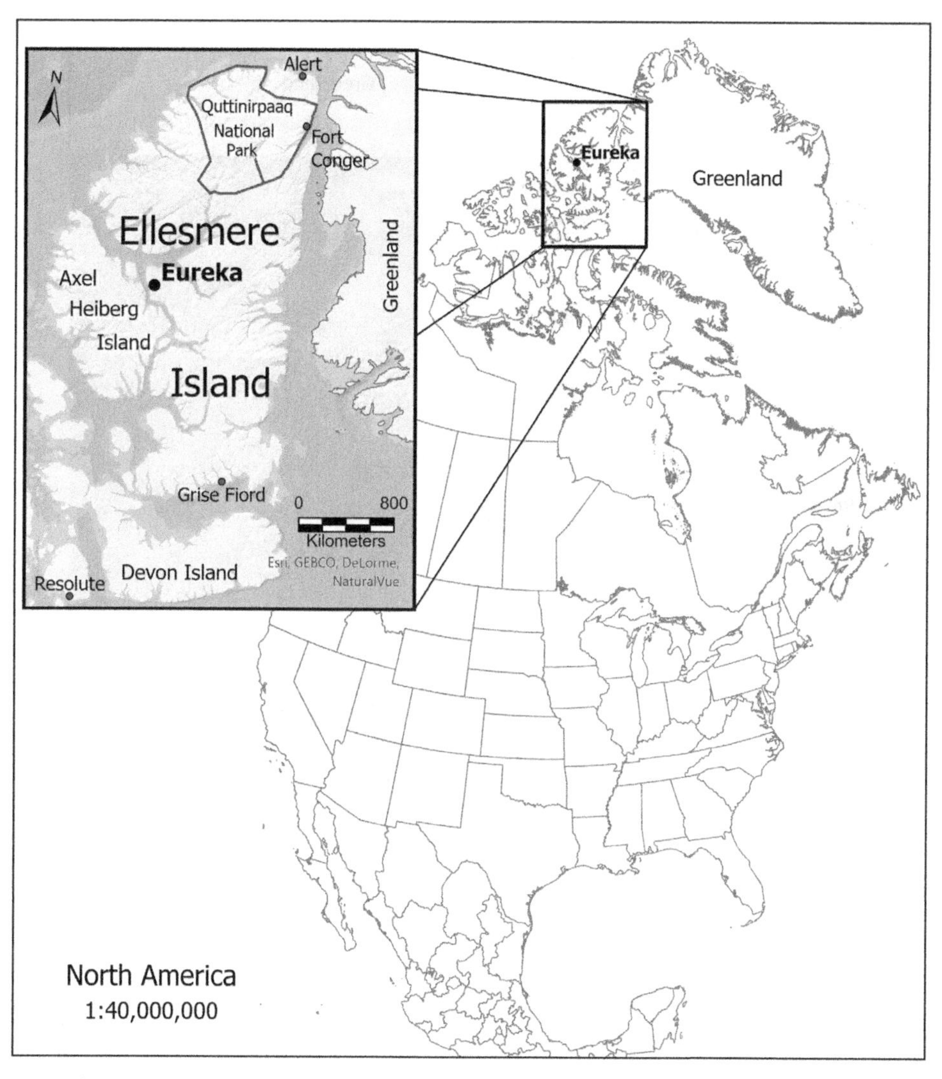

Figure 1.2. Ellesmere Island, Nunavut, Canada, is the northernmost Canadian island, lying 475 miles (764 km) from the North Pole.

hollow beneath the base of a tall, wind-carved, sandstone ridge protruding from the side of a gently sloping, tundra-carpeted valley. I had followed a breeding female to this area after trailing her off and on for a couple of days and many kilometers on my three-wheeled all-terrain vehicle (ATV). Upon first spotting her and having noted her obvious lactating nipples to feed pups, I realized that if I could stay with her long and far enough, she would lead me to her den.

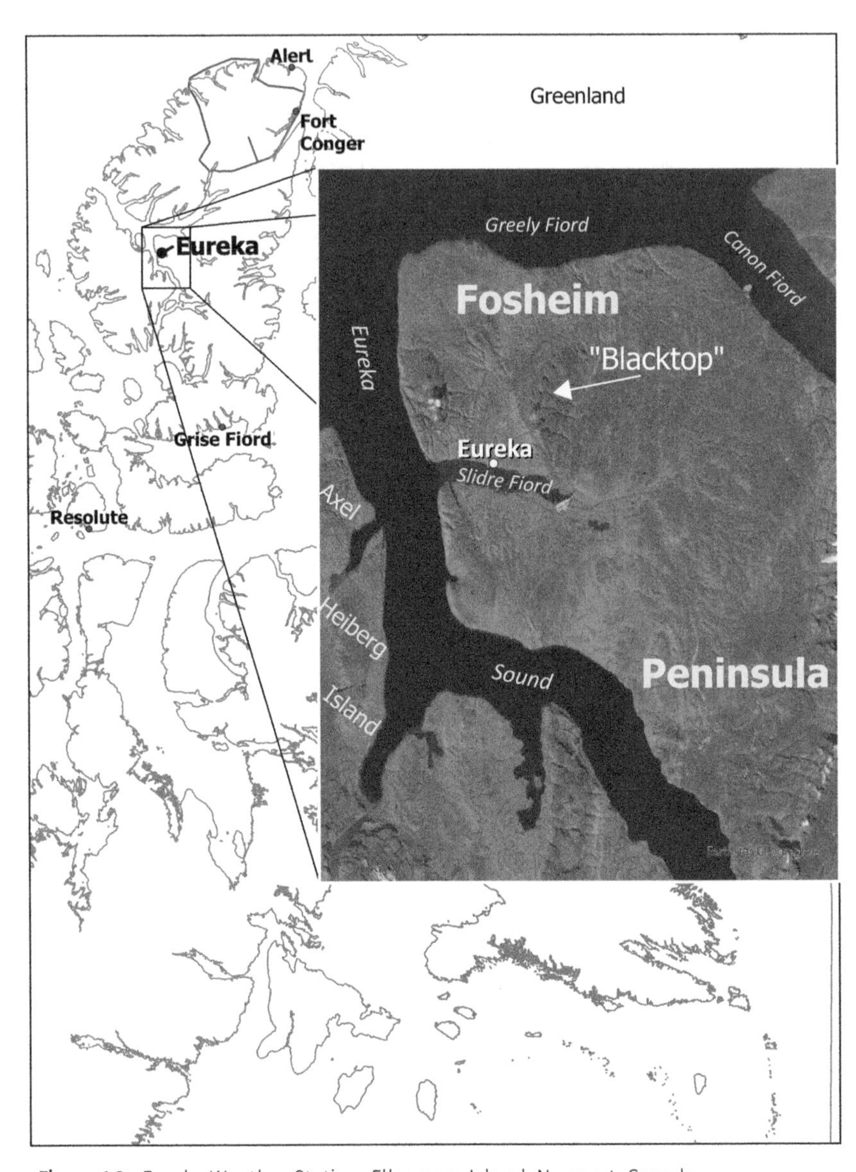

Figure 1.3. Eureka Weather Station, Ellesmere Island, Nunavut, Canada.

It turned out that it would take a false start, approaches by some other members of the wolf pack along the way, and some additional scouting around first, though. The direction the mother wolf had been heading was right toward an area where, a decade before, biologist Eric Grace (1976) had found a wolf den and had actually published a tiny map of its location. The drawing took up only a quarter of a small, scientific journal's page, so it was not precise. Still, it was a good clue now that the mother wolf had headed in that general direction.

I did the best I could to scale off the location on the map and then scout out that area, and I did finally find a hole in the ground surrounded by old, sun-bleached bones. Although the hole did not look freshly used, the fact that the mother wolf had been heading in its direction gave me some confidence. I decided to watch from a nearby hill to see if any wolves showed up.

Eventually some did, but to my great disappointment, they did not approach the possible den but rather continued past it, although still in the same direction the mother wolf had gone. If this was the den Grace had found, the wolves were no longer using it. Nevertheless, I now knew I was in the right general area but just not far enough along the right direction. Only a half kilometer or so farther is where I ultimately made my discovery.

Before running into the nursing female, I had spent days both on my three-wheeled ATV and on foot hiking over multiple hills and ridges and through various valleys, searching for tracks, fresh scats, and any other sign that might help narrow my quest for the den. Finally finding the nursing female was my first break. Now I had to play this golden opportunity just right, and I was certain I would locate the den of pups. And from the close-up view of the wolf's nipples through my binoculars, I felt the den would shelter several pups.

The female did not seem especially fearful of me on my ATV, at least from a distance, but I knew it was better to stay conservative and follow her from afar. If I pressed the wolf at all, she could outpace me, especially over rough terrain. She could nimbly navigate all that while I had to guide my tipsy three-wheeled monster over huge tundra hummocks and through rocky ground without flipping the heavy beast on top of me. That was an important consideration, given that the nearest hospital is 1,100 miles (1,800 kilometers) away.

I felt I probably knew this wolf, and that she probably knew me. A few months before, in April 1986, I had visited this area when it was covered in snow, on an assignment from *National Geographic* to write an article about Ellesmere. The island, about the size of Minnesota, is part of the Queen Elizabeth Islands cluster of the Canadian Arctic Archipelago, a region few people have ever heard of or known about, 600–1,200 miles (1,000–2,000 kilometers) northeast of Alaska and much closer to Greenland.

Ellesmere hosts a weather station and seasonal military base at a spot called Eureka along its west-central side (fig. 1.3). Besides that, only Grise Fiord, a small Inuit village at the island's southern end, and Alert, a military base at the northern tip, are the other permanent outposts. Half of the island, with rugged mountains and glaciers, remains ice and snow throughout the year; in scattered places the other half actually thaws in summer, supporting ground-hugging vegetation and a few species of mammals, most notably arctic hares, muskoxen, Peary caribou, and arctic wolves.

It was the arctic wolves, of course, that persuaded me to accept the writing assignment from *National Geographic* in 1986. The other creatures comprised the wolf's main prey. Radio-tracking wolves in a Minnesota study for the US Fish and Wildlife Service (USFWS) was my actual job, and I had to take annual leave to work on the Ellesmere article. The piece was supposed to be about Ellesmere itself, the topography, the settlements, vegetation, weather, and wildlife. It was decidedly not to be about wolves because *National Geographic* already had a wolf article pretty well ready to go that had nothing to do with Ellesmere.

Still, I knew that this was my Chance of a Lifetime. In my 1970 book I had discussed Ellesmere's arctic wolves and learned of an ornithologist, David Parmelee (1964), who had visited the Eureka area in the 1950s. Parmelee had written that the wolves in that region had so little contact with humans that the animals had never learned to be wary of them; the Ellesmere wolves were tame!

Ever since learning that, I had wanted to visit Ellesmere and hopefully see one of the white wolves. These animals, preying mainly on muskoxen and being so differently colored than most wolves anywhere else in the world, were especially intriguing.

This writing assignment was my chance to possibly see one. Ellesmere is difficult and very expensive to get to, however. The best way by commercial flights was through a little Inuit village, Resolute Bay, about 400 miles (650 kilometers) from Eureka. From there, one needed special permission to fly on Canadian government charter flights that service the weather station or various research camps throughout the High Arctic Archipelago. With my *National Geographic* assignment in hand, however, I was able to negotiate all that.

Thus, in April 1986, a photographer, two snowmobiles, and I landed at the Eureka Weather Station. Already 24 hours of daylight pervaded the area, although the temperatures hovered way below zero and the perpetual winds cut through all but the thickest layers of goose down and the latest outdoor winter wear. Those thickest layers are what I wore: padded "wind pants," and a down parka over layered pants, jackets, sweaters, long underwear, and the best outer wear. The weather station, fortunately, afforded a comfortable overnight oasis in this bitter-cold and snowy polar desert.

And that April visit was the first time I might have met "Mom" as I later called her, the wolf I had now followed to her den of pups. Back in April, the photographer and I had been allowed a haven in an old outbuilding-type bunkhouse that was part of the weather-station complex. During one of our first days there after returning from scouting out the area, I was walking to that building when suddenly seven white wolves on a ridge not many meters away caught my eye. Right there between the buildings at the weather station! Had they caught the smell of food wafting from the mess hall?

It turned out that these wolves had learned to visit the station because of refuse that was trucked about a kilometer or two to a landfill; they actually followed the truck all the way and waited while the human food remains were dumped. That was much easier than trying to run down a bunch of arctic hares or searching for some old, vulnerable muskox and trying to bring it down while avoiding its deadly hooves and recurved horns.

This situation made me wonder whether the tameness of arctic wolves was merely due to their grubbing food in such ways. However, records from various areas throughout the High Arctic suggest that wolves in general there are similarly unafraid of humans and seem curious when first meeting them. For example, in 1977, about 100 miles (160 kilometers) south of Eureka, six wolves strolled up to two botanical researchers and one jumped at one of the botanists, grazing her cheek as she flinched and jumped out of the way. The wolf then dropped to the ground, and the pack simply retreated (Munthe and Hutchinson 1978).

On Melville Island, some 400 miles (650 kilometers) away, a wolf actually licked the face of a human (Miller 1978), and elsewhere in the region a wolf fed from a person's hand (Miller 1995). Then there was Parmalee's (1964) 1955 experience of grabbing a wolf pup and carrying it in his arms while the mother wolf trailed at his footsteps many kilometers from Eureka, which I had detailed in my 1970 book. Further, in 1950, geologist J. C. Troelsen (1950) included the following in his article "Contributions to the Geology of Northwest Greenland, Ellesmere Island and Axel Heiberg Island": "Wolf: wolves and fresh wolf tracks were occasionally seen. The wolves were fearless and on a few occasions even visited my camp [in the Eureka area]." Recently, an article in the journal *Arctic* documented that these behaviors in wolves are unique to the High Arctic (Marquard-Petersen 2022).

Whatever the case, this weather-station pack of seven naturally thrilled me to no end. Just to see them was such a privilege, one I had sought ever

Figure 1.4. On one of my first days on Ellesmere Island in April 1986 I had finally seen white Arctic wolves.

since writing my 1970 book. Now, on one of my first days on the island, I had already accomplished that (fig. 1.4). But on learning that these wolves often frequented this area and seeing just how unafraid they were, my wolf-researcher wheels began spinning.

I had ten days in the area. And a snowmobile. Could I find these wolves again, trail them, and watch them while they traveled? How close would they let me get on a snowmobile? Could I perhaps even follow them from a distance, recording where they went, what they did, how far they traveled, and how fast? Could I maybe even observe them hunting muskoxen or arctic hares? Any of this would have been new and thrilling and might even allow me to learn something new about wolves in this short time.

Just being able to watch wolves from the ground was a new experience for me after spending many years viewing them as specks from aircraft. One of my biggest thrills in my long career remained the time when, as I was working on my PhD study on Isle Royale in Lake Superior, I had left the airplane. I had snowshoed to an old shack and watched from there as a pack of wolves filed by on a frozen bay, and one had stood a few meters away as I peeked out an open door at it and clicked its picture—a thrill much different than seeing a wolf far below from an airplane.

Now, here on Ellesmere in April, this situation suddenly might be a real bonanza. Over the next several days, my partner and I encountered the wolves several times both while on foot and on snowmobile. The daylight throughout the night and day helped a great deal. A giant iceberg frozen into a fiord in front of the weather station gave the wolves a place to climb on and to play. By hanging out carefully as close to the wolves as they would allow without becoming alert and disturbed, we on our snowmobiles were able to approach closer and closer.

Tossing a few pieces of leftover sausages here and there and remaining nearby with motors running as the wolves approached and fed soon acclimated the animals to the machines' roar. (Although wild-animal feeding is now banned in many areas, it was common then and, in some areas, is still being used for wildlife research; see Powell, Mansfield, and Rogers 2023.) Being able to be near the wolves so regularly was extremely satisfying and stoked my research mind endlessly. By the all-too-quick close of the frigid April visit, my partner and I could actually snowmobile along with the wolf pack, sometimes among the individual wolves.

Because we had spent most of our time habituating the wolves to us and our snowmobiles, I had not yet been able to record many new observations nor watch any hunting behavior. However, I began to eagerly look ahead to summer, when my scheduled return might eventually allow some of these dreams to come true and even result in finding an active den and observing wolf behavior around it. If, during summer, these wolves were to remain in the area and retain their lack of concern about me and my roaring machine, maybe I could travel around near them or with them on an ATV.

I knew that my magazine contract included funding for a long summer visit, a three-wheeled ATV, and even several hours of helicopter time. All of that was needed to suitably gather material from all over the island and information about its human, floral, and faunal inhabitants as well as its physical extent, geology, scenery, and so forth. By the end of my April visit, however, I already knew that my personal summer focus, while still gathering all the information I needed for the article, would be trying to find a wolf den.

: : **2** : :

The Den

And now I had found it.

It was an amazing and life-changing moment. Greeted by seven barking and howling wolves, elated that I had finally located an active den yet afraid my disturbance might cause the adults to move the pups, I quickly gathered my thoughts. I couldn't help anticipating all I might learn by observing the pack here, but I figured I should probably leave right away to minimize my disturbance even though I hadn't seen any pups yet. Still, I knew the little bundles of fur had to be in the hollow, so better to head out now and return later to allow the adults to settle down.

I had parked my ATV several meters up the hill while I strolled down the hillside, but I had carried down a pail of meat scraps from the weather station. I immediately tossed the wolves some offerings and started uphill toward my ATV. The wolves instantly scarfed up the scraps and busied themselves, noses down, searching around for more—except for one scruffy, probable one-year-old wolf (yearling), that is, that followed me up the hill. I threw it another scrap, mounted my ATV, and sped away.

"Scruffy" followed for quite a distance as I bounced over the tundra hummocks away from the den with my mind racing about all the potential I imagined now that I knew the den's location. Only a week of my summer stint here had passed, so I would have many more weeks to observe the island's wolves. At that time, I had no inkling about what kinds of observations I might eventually make or that those summer weeks I still had left here would preface so many more summers.

I would wait a day and then sneak up to where I could view the den from a different direction on foot with binoculars. Crawling up to the edge of an opposite hill, I would observe as long as I could to get an idea about how many pups the pack had and how far away I would need to stay for the wolves to tolerate me.

: : :

The chance to observe wolves while on their own terrain and watch them interact with pups came just at the right time in my career. I had captured hundreds of wolves in modified-steel foot traps, weighed the animals, taken their blood, radio-collared them, and followed and observed them from aircraft. It was now new and refreshing for me to actually be out on the ground with wolves and sharing their space while they did their thing. I had not done that since my early summer days on Isle Royale decades before. Even then, however, I only rarely ever saw a wolf, and the few times I did, I merely caught a glimpse of it.

Ellesmere brought a whole new and revolutionary experience. Already, each day had yielded different observations and portended new possibilities. How close would I be able to get to these wolves and to their den and pups? Would the adults allow me to follow them like I had been doing with Mom? From what distance? Only a few other biologists had observed wolves around a den, but in doing so, they had learned things about wolves that I did not know, the kind of information one absorbs from interacting with one's own pets—a real familiarity not afforded by a trapped wolf or from wolves observed far below an aircraft.

This was a time when other wolf biologists were helping fill the many information gaps about wolves in various other areas, often using techniques I had pioneered or taught them. Thus, researchers conducting wolf projects in several parts of both North America and Eurasia were live-trapping or darting wolves, radio-collaring them, locating them from the air, and putting location points on maps showing their travels and territorial spacing.

Collectively we had established that a wolf pack is a family of parents and offspring and that most maturing offspring disperse from the pack to find a mate and start a pack of their own. Packs raise pups in dens during spring and summer but become nomadic around large territories during the rest of the year. Parent wolves defend their territories via howling and by scent-marking with urine as well as by directly attacking trespassing wolves. Adults travel far and wide to find prey vulnerable enough to kill with the least risk, but they succeed during only a small proportion of the time, and then mainly by killing very young, old, or debilitated individuals.

However, most of the real breakthroughs in wolf biology had still not been made. The radio collars being used were standard types developed in 1963; it was not until several years after I first landed on Ellesmere that GPS radio-tracking collars were invented. Almost another decade would pass before wolves would be reintroduced into Yellowstone. And the field of molecular genetics was just emerging and had not yet been applied to wolves. Nor had I yet started my study in Denali National Park, Alaska. Even the wolf literature was pretty limited, with my 1970 book still the standard reference work.

In short, the world of wolf research was ready for a new approach, or rather the resurrection of an old approach applied to a new research situation: direct naturalistic observations recorded with pen and paper. Only Adolph Murie (1944), who had observed wolves around a den in Mount McKinley National Park, had really published much about his naturalistic findings. Two other biologists had written theses and a dissertation based at least partly on watching wolves around dens, but almost nothing in the way of peer-reviewed scientific literature had resulted from them.

In that respect, the research situation that I now found myself in had much to offer in the way of unique characteristics. Not only were the subjects of study, if not necessarily tame, then certainly far more tolerant of humans than wolves anywhere else in the world, but they also lived at the northern extreme of the worldwide wolf distribution. In fact, the closest area where wolves had been studied at that time was at least 1,250 miles (2,000 kilometers) away. On top of all that, only one or two people had ever reported on wolf interactions with muskoxen, their main prey on Ellesmere, and no one had studied wolf relations with arctic hares.

This situation was ripe for study.

And I seemed poised to cash in on it. Or at least that is what the situation so far had seemed to promise. During April, I had been able to travel along with the wolf pack on my snowmobile. On the present trip, I was able to follow Mom on my three-wheeler for long distances to the den without her running away from me. Around the den, the pack, while upset at my intrusion, had accepted my meager food offerings, and one had even followed me closely on my ATV for long distances. Now I had several weeks to test the situation, and at the very least I knew I could watch the wolves from a distance through a spotting scope or binoculars.

Ellesmere is wide open, with the nearest tree many hundreds of kilometers away. Sure, there are rocky ridges, boulder fields, hills, and canyons. And the weather is cold and windy. However, layers of clothing could solve the latter challenge, and I already had all that from my April visit. In short, the situation was excellent for long-range viewing. Nevertheless, the closer I could get to the wolves to watch, the better.

Thus, the day after I found the wolves' den, I approached the area from a hill across the valley from the den and crept to a crest about 200 meters opposite. On a sponge-rubber mattress and in full arctic garb, I gradually inched forward until the full valley in front of the den came into view. There, the whole pack lay spread out sleeping before me. Right away I counted the pups—"one, two, three, four, five, six little beige bundles," I wrote later (Mech 1987, 567)—and I judged them to be about five weeks old. I was ecstatic. Seven adults and six pups! And the best news was that the wolves had not abandoned the den despite my rude intrusion the day before.

Now, in plain view of the pack but motionless and less conspicuous, my presence would constitute a second test of their tolerance. For hours I peered at the pack through binoculars and snapped photos through my telephoto lens. In due course, the wolves started to arise, and soon the whole pack came alive in a glorious game of chase. Eventually, however, the wolves began to notice me, and for the next hour or so, various individuals including Mom and the pups approached me cautiously, with some adults coming within several meters behind me.

Without arising, I tossed my newfound companions some meat scraps, and that seemed to mollify them. What an experience! The wolves were riled up: they sometimes barked or howled, and they seemed cautious but very alert and excited. What was this strange, flat lump with weird odors protruding from the tundra not far from their den?

I had no reason to fear these creatures because when I first started studying wolves, I had even been able to chase as many as 15 of them away from a kill they had just made (Mech 1966), and just the day before, when I had found this pack's den, the whole pack had left me alone.

Suddenly my furry hat blew off, and the wolves chased and grabbed it. But the whole pack appeared quite discombobulated, and I wished I could have got out of there without further exposing myself and leave the wolves alone. Some adults circled around behind me, and to my great chagrin, Mom led her pups away from the den. Thinking I had really messed things up, I decided the best thing I could do was leave and let everything die down. Maybe the pack would return to the den. I stood up, picked up my mattress, and strode straight away, with several wolves escorting me while I threw them scraps. Glancing back, however, I did notice that a couple of adults were starting to head with the pups back toward the den.

I was now in a quandary. I was so thrilled and excited to be interacting close-up with a pack of wild wolves on their own grounds, but also I was trying to figure out just how concerned the wolves were about me. They had not run off but rather seemed more bewildered than anything else. Still, the female leading the pups away from the den was disheartening. It was critical to my plans that the pack remain at the den. Thus, should I leave the area entirely again and return a day or two later just to ensure they not abandon the den? Or should I continue to experiment, but more cautiously this time? After all, they now knew that I, an upright human being, was there, yet they seemed pretty tolerant and did return to the den.

I decided it was worthwhile to hang around a bit longer and further size things up. I waited several minutes and then headed slowly back to the crest of the hill. The whole pack had settled back down around the den again, so I settled down on my pad and watched for several more hours. It was an amazing time for me. Here I was finally just watching wild wolves around their den. Even better, I was doing so thousands of miles from civilization.

Twice more, most of the adults ventured up and checked me out, with one approaching to within a few meters. Even Mom and the pups ascended the slope toward me. But then they started heading more or less away from the den. Still, when the rest of the pack joined the little caravan, everyone headed back to the den, where all lay down and slept. I found it hard to believe, but it seemed like, despite all the disturbance I had wrought in the wolves' most sacred area, the creatures were now calm and relaxed while I lay in plain sight only a few hundred meters away. I continued to observe.

Eventually, a few adults arose, stretched, and touched noses; then five of them started to head up the hill toward me. As they approached, they looked serious as though intent on some mission. Were they going to check me out again? Or were they just going hunting? It was getting late in the day, about the time wolves often start out on a hunt. Whatever their intent, there was nothing I could do but just lie there and see what happened. I lay as still as I could as the whole pack came up to within a few meters. And then they simply bypassed me and continued on their way. They were heading out on a hunt.

I was elated. I clearly had passed the inspection of these five adults, and I grew increasingly confident that during the next several weeks I would be able to observe and study them as I had hoped. But what about Mom and the pups? They were the wolves I most wanted to be able to study. How close would they let me get without spooking? I had lots of questions to get answers to. How often do the pups nurse? When do they start depending on solid food? Who all feeds them? Is there a babysitter, or does Mom remain with them all the time?

Adolph Murie (1944, 28), a US National Park Service biologist, did a study that had hinted at answers to some of these questions but was strictly anecdotal, with no quantification. On a more sober note, there was also Murie's characterization of watching wolves around the den: "Many hours were spent watching wolves at the den and yet when I undertake to write about it there does not seem to be a great deal to relate, certainly not an amount commensurate with time spent observing these animals."

As I continued lying on my mattress on the crest of the windy hill for hours watching the den and pondering these questions, Mom and a yearling that had stayed with the pups approached me five times in the next few hours. They were still antsy, yet each time they returned to the pups. Had I passed their test too?

: : :

When I first heard of Ellesmere Island, in the late 1960s, while writing my 1970 book, I knew nothing about the place except that it was up near the North Pole. But I did know that ornithologist David Parmelee, when studying ptarmigan there, had caught a couple of wolf pups, and while he carried them back to his tent, their white, arctic wolf mother followed at his heels and acted tame. The idea of white wolves intrigued me. I had used a Parmelee photo of one of them in the book and always wanted to see one.

Although possibly seeing a white wolf was my main motivation, I also became intrigued with various other aspects of the island. To fulfill my responsibility, I would have to learn a great deal about this tenth-largest island in the world, its geology, topography, weather, human outposts, and history as well as about its vegetation, wolves and other wildlife. A budget enough to visit Ellesmere a few times for several weeks and explore it by helicopter would help me do it. How much of that time I would decide to devote to wolves would be up to me.

:: 3 ::

Ellesmere Island

The second-closest land mass to the North Pole, Ellesmere Island, lies in one of the most obscure regions on Earth. Few folks seemed to have any idea that anything much exists north of Hudson Bay or Alaska. I would later find in giving talks or discussing the island that many people thought it must be part of Alaska or near it, and I have had to explain that Ellesmere is about 1,200 miles (1,900 kilometers) from Alaska, 750 miles (1,200 kilometers) north of Hudson Bay, and much closer to Greenland.

As the tenth-largest island on Earth, Ellesmere covers about 77,000 miles^2 (197,000 kilometers^2), which is almost the size of England and Scotland combined and stretches 500 miles (800 kilometers) from north to south. Just west of west-central Ellesmere lies Axel Heiberg Island, a smaller "partner" piece that on maps fits like a complementary segment of Ellesmere.

Surrounded by fiords, many sporting huge icebergs, the island is pervaded by mountains (up to 2,600 meters high), multiple glaciers, and ice fields, some almost a kilometer thick. About half of the area remains ice and snow throughout the year. The other half, generally at lower elevations, thaws in summer and includes scattered "thermal oases" (France 1993). These areas usually lie where the sun reflects off a mountain range and heats the land.

Covering latitudes from 76°N to almost 83°N, Ellesmere endures extreme weather, with temperatures varying from –70° to 70°F (–57° to 21°C) and almost constant winds. Nevertheless, its annual precipitation of 2–4 inches (5–10 centimeters) qualifies it as a desert. The sun rises in late February, and days lengthen until the sun no longer drops below the horizon in mid-April.

Constant daylight persists until the end of August, when the sun briefly dips out of sight. The days continue to shorten rapidly until the sun no longer rises in late October.

In the ice-free areas, prostrate vegetation manages to eke out a living, thus supplying food for the collared lemmings as well as the arctic hares, Peary caribou, and muskoxen, which then support the three carnivores: ermine, arctic fox, and wolf. Various waterfowl, seabirds, and other species of birds, including the red knot, arctic tern, gyrfalcon, snowy owl, and long-tailed jaeger, nest there. The full complement of marine mammals ply the surrounding waters, including various seals, walruses, narwhals, and polar bears.

Just south of the center of Ellesmere-Axel Heiberg Island protrudes the 1,890 mile^2 (4,823 kilometer^2) Fosheim Peninsula, on which lies Eureka. The Sawtooth Mountains form a natural boundary at the southeast end of the Fosheim, and the Müller Ice Cap and Princess Margaret Range on Axel Heiberg Island lie to the west. Eureka Sound separates Axel Heiberg from the Fosheim Peninsula by about 7 miles (12 kilometers) and is frozen for most of the year, as are Greely Fiord, Cañon Fiord, and Nansen Sound to the north and east of the Fosheim.

The Fosheim region is typified by a short snow-free period between June and September, but ice is present to some extent in the sounds and fiords year-round. July, the month when I found the wolves' den there, is the warmest month, with a mean temperature of 42°F (6°C) and highs up to 70°F (21°C). February is the coldest month, with a mean temperature of –35°F (–37°C), but temperatures can drop to –67°F (–55°C). Average annual precipitation is 1.4 inches (36.3 millimeters), mostly accumulating in July and August.

Typical Ellesmere winters spectacularly belie the island's tropical past. Although difficult to fathom, Ellesmere once hosted a veritable tropical paradise, or at least a hot primeval expanse of swamp and rainforest supporting snakes, tapirs, lizards, flying lemurs, and even beavers (Plint et al. 2020) and primatomorphans (Miller et al. 2023). Camels inhabited other parts of the place. Fossil redwood forests from some 50–60 million years ago are still apparent in some areas of the island. Coal-like beds exposed by river cutbanks shed charcoal fragments that dot the mudflats along the rivers' braided channels. In 2006, a fossil fish was found on Ellesmere that is reputed to be transitional between fish and all the remaining vertebrates.

Various theories have been proposed explaining how this island so far north could have once been tropical. The leading contenders seem to be either that during geological time the island shifted to the north or that early volcanic activity produced a long-term greenhouse effect that warmed the area for many millennia.

Slidre Fiord, about 2.5 miles (4 kilometers) across, runs east and west, bisecting the western end of the Fosheim for 15 miles (25 kilometers), and was the main fiord where the wolves that I snowmobiled around with in April traveled. Eureka is located along the north shore of the fiord (fig. 1.3). The weather station there is where I headquartered during my first visit to Ellesmere and where I left most of my supplies and equipment during summers while I camped out and studied the wolves.

The Eureka Weather Station was built by the United States in 1947 but has long since been rebuilt, and is administered, by Canada, with Atmospheric Environment Services personnel stationed there year-round. In late summer, the fiord usually opens enough for a cargo ship to deliver fuel, larger equipment, nonperishable foods, and other supplies. A gravel runway suitable for large cargo planes allows infrequent but regular personnel, food, and supply flights, along with frequent military aircraft as well as various chartered planes.

A fuel cache maintained at Eureka makes this haven a must-stop for most private or chartered aircraft plying the region, including those headed to Quttinirpaaq National Park on northern Ellesmere, the North Pole, or various seasonal research camps scattered around the northern Queen Elizabeth Islands. It was by tagging along on such flights that, given my association with the National Geographic Society and later the US government, I first reached Ellesmere.

While much of Ellesmere Island is covered by mountains and ice sheets, the Fosheim Peninsula is predominantly flat to rolling, with long, rocky ridges to the northwest of Eureka running north–south. It was at the base of one of these ridges that the wolves denned in 1986 when I found them, about 3 miles (5 kilometers) north of the fiord.

The country is hilly along Slidre Fiord, including Blacktop Ridge at about 600 meters above sea level and southeast toward the Sawtooth Mountains, and is interspersed with sand dunes and open tundra. When first trying to find the wolves' den, I spent considerable time scouting some of these hills on foot and my ATV but to no avail. Creeks and rivers are plentiful on the Fosheim, but seasonal. The only lakes of note are Romulus and Eastwind, but expanses of tundra ponds are widespread during summer.

Vegetation is low-growing and sparse, generally consisting of prostrate arctic willow in the lowlands, with herb tundra to cryptogam barrens. The Eureka area is part of one of the arctic oases mentioned above. There, relatively productive vegetation extends from the Fosheim Peninsula northeast through sheltered productive valleys to the Lake Hazen Plateau (Dick 2001). Sedge meadows dominate wet areas, and great expanses of purple saxifrage spread across the higher regions, a boon to arctic hares.

Besides Eureka, only two other year-round outposts exist on the whole island. Grise Fiord, a small Inuit village (population 120), sits at the island's southern end about 250 miles (400 kilometers) from Eureka, and Alert, a military base, is at the northern tip. Quttinirpaaq National Park at the north end of the island is only visited by officials or adventurers in summer.

Grise Fiord was established in 1953 to reinforce Canada's sovereignty over Ellesmere (Qikiqtani Truth Commission 2014). The government transplanted Inuit there from Quebec over 900 miles (1,500 kilometers) to the south, stating that this was to allow the native folks to maintain their original way of life.

An Inuk, Larry Audlaluk, who was one of the first to be translocated there described what their arrival was like: "There was no type or any kind of sign of other people living up here. . . . And we only had our immediate equipment including our tents, sleeping bags, and only the little food we had with us. We didn't know where to go for additional meals after what we had was gone. We didn't even know where to go for water" (Audlaluk 2021, quoted in Bowen 2022).

Besides some meager tourist income, the community has subsisted by fishing, hunting seals, beluga whales, caribou, and muskoxen, and guiding polar bear hunts, as well as on government subsidies. During my April visit to the island, I dropped by Grise Fiord and slept overnight in an igloo built especially for the few tourists who occasionally manage to visit. Despite the hardship, the people of Grise Fiord continue to work hard to maintain their culture and traditions while forging a path forward for Inuit on their own terms.

Long before that transplant, the ancestors of the Inuit, known as the Thule, had arrived in the eastern arctic from Alaska 800–1,000 years ago. They abandoned Ellesmere in about 1700 after a long series of especially hard winters and crossed the 25 miles (40 kilometers) east to Greenland. Rock circles and collapsed piles of rocks, surrounded by scattered, whitened bones of various species of Thule quarry such as muskoxen, seals, and whales, still document their presence along the Slidre Fiord shore on the Fosheim Peninsula.

Almost 310 miles (500 kilometers) northeast of the Fosheim Peninsula, perching at the northeast tip of Ellesmere, the Canadian military base, Alert, is the world's northernmost year-round, human-occupied outpost. Established originally as a weather station in 1950, Alert is the closest North American presence to the former Soviet Union, other than across the Bering Strait, which made it useful to the military. The base once housed more than 200 troops, but its current population is about 62. The Alert area probably also hosts the northernmost wolf population in the world. (The northernmost tip of Greenland is just a bit farther north, but whether wolves inhabit that part of Greenland is not known; see Marguard-Peterson 2022.)

Around the northeast corner of the island from Alert, about 100 miles (170 shoreline kilometers) away, lies Fort Conger, one of Ellesmere's most interesting historical sites, most notably for its role with early explorers in finding the North Pole. Of course, I was interested in Fort Conger as well as Alert, but I was most intent on observing my newfound wolves around their den.

Still, I had to take time to helicopter to both sites to gather information for the article whose funding enabled me to find the den and observe around it (Mech 1988b). Fort Conger, established by A. W. Greeley on his scientific expedition there in 1873 and critical to Robert E. Peary's various attempts to reach the North Pole, remains much as it was when actually in use. Three long, low houses that Peary's crew had built from Greely's wooden fort still lie along the coastal plain, well preserved by the island's long, cold winters. Old, rusted tin cans; nails; china shards; and other detritus still decorate the area around the remains like they were discarded yesterday.

Although Ellesmere's various novel features intrigued me, by far my focus remained on the bonanza that befell me near Eureka on the Fosheim Peninsula—the wolf den attended by seven adults and six pups.

: : **4** : :

Living with the Pack

I would have been ecstatic if my discovery had merely been a den of wolves that I could view through a telescope or binoculars from half a kilometer away. It still would have been the first such den I had been able to study after all my years of snow tracking, live-trapping, radio-tracking, and aerially observing wolves and even studying them in captivity. But this den held far greater promise. The inhabitants were actually tolerating me being much closer to the den and the pups than I had ever imagined they would. I might not even need a scope.

The dilemma now was how best to make use of what clearly was an incredible opportunity. These wolves were acting almost like a pack of pet dogs. Should I try to see how close I could get to the den and the pups, as I was so tempted to do? At one point during the afternoon when I lay on my mattress across the valley from the den, I had got to within about 200 meters of it. But would trying to get closer spoil the current situation I had? Would the adults continue to tolerate me? Would they remain at the den or move somewhere else? How should I play this wonderful hand I had been dealt?

I decided to play it carefully, at least at first. I remained a few hundred meters away from the den until about 10:30 p.m., about 9 hours after having arrived. I then returned to the weather station to await the scheduled arrival of the photographer, who I knew would be especially pleased with how tolerant the wolves were. *National Geographic* magazine is famous for its photographs, and although wolves were to make up only a mention in my article about the island, I knew that a spectacular picture of a wild, white wolf, or perhaps a batch of pups, would make an attractive candidate for inclusion in the piece. And I, a mere amateur photographer myself, had already snapped quite a few good shots of the wolves as they approached me, so a professional should surely find this wolf situation a total treasure.

Just as wild wolves had always been very difficult subjects for biologists to study up close, so too were they very difficult for photographers to deal with. This was 1986, long before digital cameras existed and adapters to attach them to 60-power telescopes, remote cameras, and drones. Nor had wolves yet been reintroduced to Yellowstone National Park, where currently park visitors photograph them virtually every day. Hence, good photos of wild wolves were quite rare. Most wolves, being so shy of humans, and inhabiting mostly remote, poorly accessible areas were a real challenge to photograph.

Here, finally, almost at the end of the Earth, existed an amazing photographic opportunity. Thus, if these wolves I had just shared the afternoon with continued to cooperate, they could well become stars in unique scenes around the photogenic den. Similarly, and more important to me, they could well become subjects of scientific publications about my behavioral studies.

What I did not realize at the time was that for both photography and behavioral investigations, these wolves would soon provide far more unique and amazing opportunities than I ever imagined.

The value of human-tolerant wolves to the photographer as well as the ability to get close to my subjects meshed precisely with my needs to study the interactions between individual wolves and between the adults and pups. Thus, in terms of how far from the den we should camp and how best to interact with the wolves, we found ourselves in total agreement on responding to what potentially would be a lifetime opportunity for both.

We two ebullient men on our ATVs headed straight out to the den area and found that the wolves were still tolerant, even of a pair of humans. The photographer was amazed, having hardly been able to believe my account of my previous day. We scouted out a campsite in the general vicinity of the den but just out of sight and several hundred meters away so that the wolves did not seem disturbed in the slightest. By merely strolling a few meters around the bend in the flats where our tents were set up, we could actually see the den ridge and often some of the adult wolves, their white pelage contrasting nicely with the extensive length of beige rock on which they lay around the den.

The den itself was a long cave-like tunnel at the base of a tall sandstone abutment at the end of a rocky ridge. The side of the ridge facing the campsite slanted up out of the terrain to reach a 2–3-meter high, miniescarpment on the opposite side. Part of the top of the ridge was level enough that the adult wolves could walk or lie on top. We later learned in viewing the den entrance head-on that the beige sandstone monolith rising above the den entrance had for millennia been weathered and wind-carved. Various lighter-colored layers protruded just enough to decorate it for great *National Geographic* photos.

In fact, years later, long after my article depicting that den appeared in the magazine (Mech 1987), a biologist, who had been there a decade before me, recognized it. Eric Grace, the author who had published the 1976 matchbook-sized map that I tried to use to find the den, wrote me that this was the den he had found and tried to map the location of. The old den I had first discovered by following Grace's map turned out not to be the den Grace had located.

Now I had information that this den had been used at least off and on for over a decade, not surprising because there appeared to be few other such protected caves possible in this permafrost-covered area. In fact, I later learned, based on radio-carbon dating, that an old wolf-cracked bone I found around the den was left there between 583 and 983 years before, and another, 162 to 302 years before (Mech and Packard 1990). How many generations of wolves had possibly used this den?

I had been focused on finding an active wolf den for so long because the den is the headquarters of a breeding wolf pack where they can safely raise their pups for their first couple of months of life. With the pups being always there, one can be certain to find them, and at least one adult there at almost any time of the day during that denning period and then observe and study them, just as Adolph Murie, and later Gordon Haber (1977), had done with a den in Mount McKinley National Park (now Denali National Park). Those studies, however, had to be done from a distance, whereas it now seemed I could study the Ellesmere wolves from much closer and thus could gather more detailed information.

Another extremely beneficial factor in this extreme-north study area was the constant daylight. All summer, the sun remained above the horizon, making a huge 24-hour circle without ever disappearing except when clouds obscured it. Thus, whenever the wolves were doing something new or interesting, day or night, one could observe and record it. And photograph it.

What I still didn't realize, however, was that quite aside from all the wolf behavior I could watch around the den area, if I could stay with the adults as they traveled and hunted, I could also watch them closely while they spent their lives away from the den during their normal daily activities. Although I had watched wolves from small aircraft for many hours, actually being with them on the ground as they headed around their huge territory seeking and capturing prey would give me far better insights into their daily lives. And more exciting ones, too.

It was the three-wheeled ATVs that would make that difference. New to me, these machines seemed to have been made specifically to negotiate the Ellesmere terrain, at least most of it. Over extensive barren ground, gravel fields, and even braided river bottoms, they could more than keep up with the wolves. Even over many of the smaller hummocks, they easily stayed with the creatures. Only large hummocks, boggy areas, deep gorges, or boulder fields would slow or stop the machines, or if not the machines, the riders. I hadn't anticipated the possibility of using these handy conveyances to travel along with the wolves just as I did in April with the snowmobiles.

In fact, as I soon learned, my ATV also made the difference as to just how tolerant the wolves would be to me under varied conditions. True, I had found that even without the machine, the wolves seemed to accept me, at least while I remained flat or low to the ground. However, as my companion and I began working around the wolves, observing and photographing them, it became quite apparent that to be completely accepted by all the pack members as well as the pups, it was best to remain atop the ATVs.

My notes from July 15, 1986, just three days after I had lain on my mattress across from the den and had wolves swarming around me, illustrate the situation: "We sat on the trikes till about 1:00 p.m. and the wolves continued in their complacency. However, we then got away from the trikes and started walking E. As we got about 20 feet away, the adults began looking up from their sleep, and the farther from the trikes we got the more alert and excited the wolves got." When we were about 60–70 meters from the ATVs and began heading back to them, the breeding female led the pups out of sight.

The other important factor that reinforced the wolves' tolerance was being fed. I had learned from my April visit that occasionally feeding the wolves greatly helped me gain their confidence. It was already clear that weather-station personnel had been feeding them, so I not only felt no qualms about doing so but also realized how advantageous it would be to my studies if doing so would allow me to find their den and to promote their tolerance of me. In April, when I was first feeling my way around the wolves and gauging how they responded to me, I fed them more often, although hardly enough to satisfy their daily requirements. I wanted them to still need to hunt naturally.

Now that I had found the den and again gained the trust of my study subjects, I decided to restrict my feeding to smaller amounts mainly to continue reinforcing their tolerance or gaining the confidence of the pups. The wolves were so eager to snarf up any little tasty morsel that it seemed small pieces were all that were necessary, although I did sometimes treat them with larger chunks. Over the summer, I learned that small fragments of arctic hare were favored, but even cheese bits or pieces of sandwich were acceptable.

Thus, I kept a supply of those foods in a closed plastic container attached to the ATV handlebars and ready to offer if the occasion warranted. Only many years later would I discover the extreme power this approach had over the wolves (Mech 2017a).

I would also eventually learn something about the wolves' mental abilities related to food. The adults seem to have an excellent recollection of where each piece of food lands. Thus, if they eat a closer one first, they remember later that another piece or two landed elsewhere and they then search for it. I even tested this idea by throwing three pieces of dry dog food in three directions very quickly: (1) toward the wolf, (2) right, and (3) left. The animal ate the closest first, then searched for the other two (2–4 meters away). She got the right one right away and searched for the left one but did not find it. However about 30 minutes later she came back, looked for it, and found it.

By reinforcing the wolves with tidbits every now and then and remaining on our ATVs the photographer and I became fully accepted by both the adults and the pups and settled in every day around the den to observe and photograph our canid companions. We actually felt part of the pack, remaining within about 15 meters from the den itself, with growing pups frequently visiting us up close or exploring around us and adults paying no mind to those interactions.

In fact, the familiarity between us and the wolves almost got out of hand when the wolves turned the tables by visiting our abode a few hundred meters from the den. With three small tents, a few miscellaneous metal cases, plastic water jugs, and aluminum cookware smelling of food, that area proved to be a strong wolf attractant. Eventually the adults and even the pups found our camp irresistible, and we had to chase them away, hooting and hollering as we did. The trouble is, I wanted the wolves to fear the camp but not us, at least not us on ATVs.

The solution was for the us to dismount from our ATVs when at camp and chase the wolves on foot, and that worked. It took a few lessons, including a major one when a bold wolf stuck its head through a round port in the back of my tent while its associates gathered around expectantly. Suddenly the wolf withdrew its head while towing my red sleeping bag, greatly exciting the bystanders. I let out the loudest, sharpest hoot I could, and the wolves dropped their trophy and fled for several hundred meters.

Finally, an uneasy truce was reached whereby we and the wolves could frequent each other's sacred ground so long as we remained on our machines, and the wolves refrained from messing with any of our equipment.

Thus, we spent the summer basking in their glory, with the dreams we both had nurtured for decades actually coming true virtually every day. We learned to identify each wolf based on coat characteristics, old wounds, and various other peculiarities seen close-up. I took copious notes about the wolves' interactions with each other and with the pups around the den, and the photographer took thousands of photos. We watched as the adults departed daily on their hunts and as individuals returned with prey parts and full stomachs to regurgitate to the pups and the breeding female, the usual way of feeding them

We soon realized that we could even accompany the wolves on their hunts, although that meant leaving the pups for a while as well as any adults that remained to babysit. The situation was truly an abundance of riches for any wolf aficionado, which both of us had been for so long. Stay near the den and observe and photograph the pups, or travel with, observe, and photograph the adults? Decisions, decisions!

Of course, we did both, alternating as the situation called for. We were both there to gather material for our article about Ellesmere Island. However, we both soon realized we had plenty of wolf material for that after just a few days with our subjects, especially because wolves were only to be a small part of the piece. Still, we knew that potentially we could gather much more exciting wolf-related data and photos than merely details of life around a wolf den. That would certainly do, but what if we could document some good hunts?

Furthermore, it soon began dawning on us that, with our newfound bonanza, there was far more we could accomplish than merely gathering material for a small part of an article, *National Geographic* or otherwise. Other articles? A book? A TV documentary? The latter would require another summer there, but, of course, that was on both our minds. How could anyone in our position with such a goldmine of material not return another year? In fact, doing a TV documentary could guarantee funding for another summer's expedition to this remote area. It was a natural fit.

And how could we assure our dream of securing funding for a documentary and thus future visits? By documenting a successful muskox hunt. Regardless of any future media prospects, documenting a successful muskox hunt this first year, or even an unsuccessful one, would not only royally fulfill our needs for articles but would also greatly satisfy us personally.

I well knew that if we were to follow the wolves enough times, sooner or later we would be near them when they made a kill. After all, each day we had watched them bring home food for the female and pups. Often these "catch and carries" consisted of arctic hares or parts thereof, but muskoxen were also on the menu. I had many times followed wolves by airplane on Isle Royale and in Minnesota and got to see them hunt, but I grew eager to watch it all from the ground, at wolf-level. Still, here I found it so rewarding to wait around the den and watch the pups that it would be a while before I could get myself to gamble on traveling away from there with most of the pack and losing any information about the goings-on around the den.

As luck would have it, I didn't need to travel far. Early one afternoon not long after we had settled into a routine with our wolf watching, a herd of 11 dark-brown, adult muskoxen and three calves approached a broad valley visible from the den a couple kilometers away. Would the wolves spot them too and head after them? We could see the dark animals clearly with our naked eyes, so could the wolves also? Or would they perhaps smell or hear the burly beasts? The pack had been starting its hunting forays in late afternoon after resting and sleeping most of the day. So, too, the animals did on this day. Thus, I wasn't surprised when the wolves did not seem to notice, or at least had not yet shown any interest in, the approaching herd.

To be ready in case they did, we decided to drive to the herd and hang around it. Then, if the wolves did spot the prey, we would be able to document any interaction right from the beginning, rather than try to keep up with the wolves. This technique had served me well during my first wolf study decades before. In Isle Royale National Park, where I did my PhD research following wolves with a light aircraft during winter, I had learned to direct my pilot to fly ahead of the traveling wolves to spot any moose before the wolves did. Then I could record the entire interaction starting from when the wolves first detected the moose right through to the final outcome. In total, I ended up watching those wolves interact this way with 131 moose during three winters (Mech 1966).

Now, on Ellesmere, we perched on our ATVs watching the nearby muskoxen methodically graze their way along the tundra while the white wolves, visible as light specks against the dark background around the den, continued to snooze away the afternoon. Eventually, however, some internal alarms aroused the hungry carnivores pretty much on time in early evening, and all seven threaded their way single file into the scene. Meanwhile, we atop our ATVs, cameras ready with telephoto lenses some 150 meters from the herd and off to one side, were as tense as we imagined the muskoxen to be as these clumsy creatures shuffled their way in front of their calves.

Would the wolves suddenly just charge the herd, or would the closing crafty canids merely size it up from afar and apply some clever strategy to pierce that shaggy protective wall?

I already knew the pack would target the calves, for that was the wolves' standard operating procedure with every other prey during summer—go for the young, weaker, less dangerous ones. Still, there was that massive brutish barrier to overcome in this case, a dark wall of broad-faced, hook-horned beasts ever ready to meet them head-on (fig. 4.1). A kick from a hoof could end it for a wolf, and many have been injured or killed by such a blow (Pasitchniak-Arts, Taylor, and Mech 1986).

As the wolves approached the muskox herd, the adults huddled together facing outward, rumps touching and rubbing, obscuring the calves behind. The herd took the situation seriously, but the wolves seemingly not so much.

Figure 4.1. As Ellesmere wolves try to target muskox calves, they are usually faced with a dark wall of broad-faced, hook-horned adults ever ready to meet them head on.

They moved in casually, at least at first, possibly trying to figure things out. Dinner was there and waiting; they just had to decide how best to get it without getting killed. They did not appear to be well organized, however, as some lay down while others wandered about, a few within meters of the herd.

Maybe these wolves and muskoxen even knew each other. After all, both groups have no doubt lived in the same area for years, and when one species was the natural meal of the other, they could have interacted hundreds of times. If so, the strategy, tactics, strengths, and weaknesses of each would have been familiar to the other. In any case, both predator and prey took quite a while to size each other up.

Individual wolves strolled between some of the muskoxen, while various members of the herd shifted and shuffled, ever maneuvering to protect the three calves—their group's total reproductive investment for the year. An individual muskox would now and then charge a wolf, which then would back off, but most of the herd tried to keep tightly to the group. Light skirmishing continued for several minutes.

But gradually the wolves grew more and more excited, and the pace of the struggle picked up.

One of the big advantages the wolves possessed in this situation was their flexible feet. The tundra there was a mat of scattered hummocks the size of the muskoxen themselves but broken up by long slots deep enough to catch a lower leg. And muskox legs are spindly, terminating in blocky hooves that could do a number on any wolf they struck but that also could catch in the tundra troughs. The wolves, on the other hand, were nimble enough, with spreading toes easily grasping the sod, as the canids twisted and turned while worrying the muskoxen into veering and fleeing.

Excerpts from my July 15, 1986, field notes captured the frenzy of it all:

> The wolves would run right into the herd, and the muskoxen would thrash about and turn on them while running, and the wolves would turn about and the chase became one big swirling mass of dark muskoxen and white wolves.
>
> After about ½ hour of that, during which the muskoxen stood their ground about 20 times, and the wolves got them running again, the herd ran to the top of a low ridge and then back to its base.
>
> Within seconds, a couple of wolves grabbed a calf out of the middle of the herd and the cow turned and fled while 2, 3, and 4 wolves grabbed the calf and tried for 5–10 min to pull it down.

Meanwhile, one or two of the wolves suddenly left the calf and continued after the rest of the herd to the north, crossing a river. The herd by now was in a complete rout or stampede and the wolves grabbed a second calf in the river, although I did not directly observe that kill. The herd turned southeast and fled along the river, then up a hill, and two, three, and four wolves grabbed the third calf two thirds of the way up the hill. Three calves! That was that little herd's whole, annual, offspring production soon to be converted into wolf pups.

I was astounded at the manner in which these wolves, already so tame, tolerant, and almost friendly to us, suddenly switched into their predatory mode and bore down on their prey. Not that this really should have been surprising. However, to a biologist who had previously only watched from an aircraft antiseptically while wolves charged after their prey and tore into it, the essential rawness of the process came bloodily into focus now that I was almost part of the pack while they procured their next meals "on the hoof."

My overall impressions of this sudden turbulence that had broken the still, calm tundra tableau spread out before me and featuring "my" wolf pack in primal action were memorialized in my book *The Arctic Wolf: Living with the Pack* (1988a) and in my July 15, 1986, field notes:

> I was really impressed with the following: (1) How automatic and frenzied the chase and killing became when the herd stampeded. The wolves "knew" when a kill was possible, and their mood changed completely. They went into high gear. (2) The amount of caching. (3) The degree of secretiveness about caching. Each animal goes off by itself and watches to see that no other individual is nearby. They seem almost paranoid about the caching. (4) How, once the kills were made, the wolves became so "businesslike" and swung into action. They reminded me of firemen at the scene of a fire, and the conversion from sitting around to suddenly putting out a fire was similar.

Now, after only ten days this summer on Ellesmere and having already witnessed the ultimate scene—wolves killing muskoxen—my elation was unbounded. What more could I ever ask for in the way of observing wolves up close? This was truly a wolf biologist's dream come true.

: : **5** : :

The Plot Thickens

My mental wheels had been spinning furiously over the last few days as I increasingly realized how truly unique this opportunity was for learning about wolves in a way that my previous decades of study had never come close to. Tracking and observing wolves from an aircraft, live-trapping and examining them, blood-sampling and radio-tracking them, tracking them in the snow on foot, observing and doing physiological tests on captive wolves had all helped turn up new information, true. But none of those studies had allowed anything like this close-up, real-life familiarity of actually living with these objects of my research attention.

I was now gaining the kind of personal knowledge of these wild creatures that one absorbs about their pets by simply living with them, and that was probably the most valuable information I would gather during my 1986 stay with the wolf pack. The potential for doing all kinds of scientific studies of these human-tolerant wolves was clear, but much of what I was already learning would not be the kind of information subjected to analysis and statistical treatment. Rather, it would be a more intimate kind of knowledge and familiarity that would even be difficult to convey to other people. How can a person truly inform someone else how their little Fido reacts when, for example, the kids get sick? On top of that was the extremely valuable, personal satisfaction of sharing daily in the everyday existence of one's lifetime research subject.

All this added up to my keen, growing desire to both ensure that all this wondrous experience would not end with the summer and also that I would be able to share it with others. How that could happen was yet to be determined, but the potential of this situation was so evident that surely there would be ways. The fact that the current trip was supported by *National Geographic* magazine meant that these early findings would get at least some publicity, which should help foster additional support.

It did. For a total of 24 summers! And then some.

Contrary to the magazine's original plan, the upshot of my summer's work was not to be an article about Ellesmere Island itself, although that would come later (Mech 1988b). Rather, the discovery of these wolves was to become the sole subject of the article, and not just any article, but the magazine's May 1987 cover story (Mech 1987). Furthermore, when the folks at National Geographic saw the material, it was axiomatic that the next product about these wolves was to be a National Geographic TV documentary, which then furnished funding and new, safer ATVs, four-wheelers this time, for a second summer's work.

The main emphasis, then, for the effort in 1987, although secondary to obtaining more basic information by living with the pack and recording the behavior of its members, was to cooperate with the cinematographers. They were both with National Geographic and the British Broadcasting Company (BBC), a partnering that worked well and yielded spectacular footage that culminated in the 1988 National Geographic Explorer documentary *White Wolf*. That film was aired several times on national TV, became available on video tape and eventually as a DVD, and excerpts are currently on YouTube at https://www.youtube.com/watch?v=6RbFtgC5TVI.

Since *White Wolf* aired, two additional documentaries were made during my tenure on Ellesmere, one done by Jeff Turner of Turner Productions and another featuring British actor Timothy Dalton. Both men spent a few weeks on Ellesmere with me, and Dalton also accompanied me on my project in Minnesota. After 2010, my last summer on Ellesmere, five more documentaries were then made there. Although they failed to mention my work, there can be little doubt that our earlier articles and documentaries laid the critical groundwork for their productions. The producers of the first few consulted me regarding the horrendous logistics of trying to work in the area, and I did my best to help them, even encouraging them to retain my colleagues to guide them. That worked out well for both the documentarians and for my colleagues, who then were able to continue our studies there. In 2012, PBS produced *White Falcon, White Wolf* (https://www.pbs.org/wnet/nature/white-falcon-white-wolf-introduction/3323/). The BBC then followed two years later with *Snow Wolf Family and Me* in three episodes (https://www.bbc.co.uk/programmes/b04ww480). Next came "White Wolves: Ghosts of the

Arctic" in 2017 for CBC's *The Nature of Things* (https://www.youtube.com/watch?v=s0QFE-hxgI8). Then, in 2018, National Geographic made excellent use of drones and remote set cameras for three episodes of *Kingdom of the White Wolf* (https://www.natgeotv.com/za/shows/nationalgeographicwild/kingdom-of-the-white-wolf). As recently as December 1, 2023, *Planet Earth III behind the Scenes* aired "Up Close and Personal with a Pack of Arctic Wolves" (https://www.youtube.com/watch?v=IXj8S0p9STw).

The National Geographic Society also had a research funding arm, and I tapped that for another five summers' trips. As a wildlife research biologist studying wolves for the US Department of the Interior, I was taking annual leave from my regular job for my special Ellesmere trips. Eventually, however, my supervisors realized that the Ellesmere project featured such a superb situation and was yielding such a rare and fruitful trove of information that they suggested I incorporate that study into my regular research duties.

From then on, my own project supported my annual summer trips to the island right through my last summer's visit in 2010. Only during summer 1999 did I fail to make the trip when, in Resolute Bay on Cornwallis Island, a few hundred kilometers south of Ellesmere, my back went out, forcing me to return to the States for the summer and forsake my annual Ellesmere foray.

In addition to the Ellesmere wolves being such cooperative subjects of this unique research situation, the local logistical set up was also ideal. Ellesmere being so remote, desolate, and isolated, most such areas there present very difficult challenges for any sustained fieldwork. However, only about 8 kilometers from the wolf den lay Eureka, a veritable—although tiny—outpost of civilization, housing Environment Canada's weather station and a summer military base, along with an airstrip. During the earliest days of my stay there, a telephone maintained by the military base was even available part time. Eventually, the internet even showed up, and email became usable.

I camped near the den each summer, or during years of no denning, near where I could see the Eureka landfill, the most likely place for the wolves to visit. Thus, the weather station became a place where I could store my camping equipment, personal gear, food, and three ATVs accumulated during my first few years. I could also obtain gasoline there for the machines. I soon appropriated part of an old, abandoned weather port there for overwinter storage. Weather-station personnel were mellow folks who were interested in my research and welcomed my presence. Mechanics there could be hired to maintain and repair the ATVs, and the guys eventually built me a special storage shack for them.

After a few years, it became apparent that the appropriate approach to setting up an ongoing research project there was via Canada's Polar Continental Shelf Program (PCSP). Logistics throughout the High Arctic were so horrendous that the Canadian government had instituted PCSP to facilitate research there. PCSP had a major base at Resolute Bay, an Inuit village of about 230 people on Nunavut's Cornwallis Island, 400 miles (660 kilometers) from Eureka. The agency also often housed one or two folks in a shack at Eureka.

By applying to PCSP in October each year, an approved US researcher could hire PCSP flights from Resolute Bay to and from Eureka the following summer. Commercial flights a few times each week connected Resolute Bay with Yellowknife in the Northwest Territories, and in some years from Montreal. Total costs for flights from the States to Eureka amounted to $8,000 to $10,000 each year one-way.

Additional flights to Eureka included the weather station's resupply flights every three weeks during summer and occasional flights for construction crews, tourists to the island's national park, and miscellaneous North Pole adventurers. Any flight coming anywhere near Eureka would stop there to refuel because of the Eureka airstrip's aviation-fuel cache, which was delivered late each summer by a barge that also supplied bulk goods and any heavy equipment the weather station needed. The nearest other available fuel was 350 miles (560 kilometers) away at Thule, Greenland, or at Resolute Bay, although individual research projects sometimes cached aviation fuel in their own handy locations.

All of these logistical advantages of conducting studies near Eureka greatly facilitated my research, but one factor lacking was reliable two-way portable-radio communication. There was no good reliable way to communicate with the weather station in case of some emergency. Thus, it was important for me to have a partner most of the time for any extended trips to Ellesmere, although I did spend a week in 1986 and in 1993 without one. Needing a companion meant funding had to be available for that second person.

During the five years of National Geographic research funding, that grant covered it, so I chose partners who had studied wolves, assisted with such studies, or in other ways were associated with my work or with the International Wolf Center, a nongovernmental organization I had founded just before the Ellesmere work. It was my way of making sure other people who could really appreciate the unique Ellesmere opportunity had a chance to do so. When I was funded out of my own government research, this support did not include salaries or travel for partners. I then had to select companions who could find their own funding, but there was always someone willing and able.

My top priority for companions was fellow wolf biologists with whom I couldn't wait to share the unique opportunity, so the following joined me at one time or another: Layne Adams, Luigi Boitani, Diane Boyd, Dean Cluff, Fred Harrington, Dan MacNulty, Jane Packard, Rolf Peterson, and Bob Ream. Wildlife technicians on my projects or other scientists included Nora Gedguadas, Jay Hutchinson, Mary Maule, and Una Swain. The following International Wolf Center board members or staff also helped: Nancy Gibson and husband Ron Sternal, Cornelia "Neil" Hutt, Walter Medwid, Mary Ortiz, Jerry Sanders, and Ted Spaulding. Others who assisted me were Greg Breining, outdoor writer; Craig Johnson, US Fish and Wildlife Service; Valerie Gates, philanthropist; Tom Lebovsky, book publisher; and Jeff Turner, cinematographer. My wife, Laurie, accompanied me there twice.

I had quickly developed a working routine around the wolves that seemed to minimize frightening them and thus maximize my ability to observe and interact with them close-up, including taking notes, photos, and videos, eating lunch around them, and traveling with them when they went hunting. Through the years, this routine became second nature to me. When I or one of my partners accidentally broke the routine, the wolves quickly reacted, retreating or leading the pups away. A usual way for an adult to do this was to pick up a bone and head off; the pups were sure to follow.

I had quickly learned through regular daily interactions with my sensitive hosts that suddenly scaring them sometimes had lasting effects. Some, if they spotted me off my ATV, grew jumpy, so I limited such occasions to when they were out of sight. At times, even changing my outer wear freaked out certain of them. A standard response by wolves skeptical of me was to hide behind a rock or ridge and peek at me, that is, barely lifting their head above the object such that only their eyes and ears showed. By maintaining a standard routine, however, I was able to retain a close and trusted relationship with the pack.

Still, each new person who accompanied me had to be taught the routine. Most important was to remain on the ATV as much as possible and to keep all gear off the ground and attached to the ATV with bungee cords. Stretching, walking, or urinating were to be done on the side of the ATV away from the wolves and, whenever possible, were restricted to the area immediately next to the machines. Around camp, miscellaneous objects were to be kept attached to the ATVs or in the tents, and food was stored in wolf-proof metal boxes.

Generally, the summer period during which I visited Ellesmere ran from about mid-June through early-to-late August, although specific periods varied considerably (table I.1). It turned out that, depending on the weather, due to water not being able to penetrate the permafrost, the tundra in June might be so muddy after the snow melted that it was impossible to walk or drive ATVs there without getting totally weighted down by mud. Although it could snow there anytime, by August, the chance of snow greatly increased.

And I did see my share of snow during my stays. An especially memorable snowstorm occurred on July 1, 1996. I wrote the following:

> Facing the very strong (prob. 50 mph) wind and light snow, we [Rolf Peterson and I] observed for about 1 1/2 hours but saw little. As it snowed and blew ever harder, we called it quits and headed to the tents to ride out the weather. It was hard at times to steer the ATV's because of the strong winds. When we got to camp, we found 2 of the tents collapsed and the third close to collapse. All we could do was put rocks on them to keep them from ripping. We also collapsed the third tent ourselves and weighted that down. We later learned that at the weather station, the winds hit 40 mph, and since we were quite a bit higher than the WS, it probably hit 50. We decided to try to ride out the wind so we could put the tents back up. Thus, we cooked some food and then huddled in the lee of a rock pile, the only natural shelter around.
>
> We remained huddled there for several hours.

An important complication to my Ellesmere routine was the fact that over the long period of the Ellesmere studies, I also had responsibility for other important research. My study on wolves and white-tailed deer in Minnesota's Superior National Forest, which started in 1966, was ongoing, complete with a full-time PhD assistant, graduate students, and technicians. Also in 1986, I was asked to head up a wolf study in Denali National Park, Alaska. I hired two Minnesota wolf biologists to conduct the project but needed to spend time there myself. Then, in winter 1995 and 1996, I helped procure wolves in Canada for the Yellowstone wolf reintroduction and began advising a series of graduate students associated with research in that park. I also chaired the IUCN (International Union for the Conservation of Nature) Wolf Specialist Group and was vice chair of the International Wolf Center, which I founded in 1985.

Because those projects all needed frequent tending, I often had to flit from Ellesmere back to Minnesota for a week or so and then head back to the island. While I was away, my companion needed a partner for safety's sake, so that meant recruiting and training additional people for the project. Nevertheless, each person worked out well, and of course was elated by the opportunity.

: : **6** : :

A Flourishing Family Falters

Over the 25-year period that I worked with the Ellesmere wolves, the research situation varied considerably. During 1986, 1987, and 1988, the wolves denned where I had first found them. In 1986 and 1987, all seven adults appeared to be the same, along with six pups in 1986 and five in 1987. Thus, for those two summers, my main objective was to get to know each wolf individually and determine their status and role in the pack and how each interacted with each other and with the pups. I considered any nonpup an adult because pups grew to adult size by December each year. Thus, except for the breeding pair, most adult-sized individuals were yearlings, that is, born the previous year.

By 1988, some of the individual adults "turned over"; that is, some remained, others left, the previous year's pups became yearlings, and some yearlings stayed with the pack. In 1988 there were four adults and four pups, with an occasional fifth adult. The breeding female was the same as in 1986 and 1987, but the breeding male turned out to be one of the other adult males from those years, an individual with a scar on his left shoulder, which is what I ended up naming him. Along with them were two other unknown individuals that were no doubt yearlings (Mech 1995b) because they were not present as adults earlier, did not scent-mark, were submissive to the breeding pair, and were responsive to my feeding routine from previous years. One was a female much whiter than any of the other adults, which I called "Whitey," and another had quite a bit of gray on his back and was thus dubbed "Grayback."

The main focus of research in 1988 was a study of the pups nursing and their interactions with the mother as she weaned them (Packard et al. 1992), as well as regurgitative food transfer between the adults and between the adult and pups (Mech et al. 1999). More about these studies later.

This variation that began in 1988 complicated trying to identify individuals. During any specific summer, I could usually keep track of the different patterns of pelage shedding on individuals, except for the pups, which could seldom be distinguished as individuals. However, to identify individual adults from year to year, one needed more permanent markings like scars, ear tears, broken teeth, and similar characteristics. I even tried to see if whisker-hair patterns might be distinctive, like in some species, but they were not. Thus, there were cases where I could not be certain of some wolves' identity.

What ever happened to the six pups in the 1986 litter or the remaining two from the 1987 litter remained unknown, along with just about everything else about these wolves during September through May. That problem really bothered me; I yearned to know what these wolves that I got to know so intimately during summer did during the rest of the year. Did they live in the same area or head elsewhere? South of Ellesmere, wolves are migratory, following migratory caribou herds for hundreds of kilometers between calving areas and wintering grounds. However, Ellesmere wolves' prey were only nomadic so far as anyone knew. At least they were not known to migrate.

But how far and wide did my wolves have to hunt during fall, winter, and spring to find their prey and sustain themselves? Thirteen wolves (1986) need a lot to eat, generally at least 3.25 kilograms/wolf/day, or 100 pounds (45 kilograms) for the pack every single day. In my other study areas, I had radio-collared wolves, which allowed me to understand how they fared year-round. But not on Ellesmere. Thus, the idea of radio-collaring the Ellesmere wolves was ever on my mind. During those first few years of the study, however, radio-tracking technology was so basic that one would have to be on Ellesmere to track the radioed wolves throughout the year, which was out of the question.

I also knew that new types of radio-collars were being developed using the Global Positioning System (GPS) and that they would be able to send location data year-round to websites. In fact, my Denali project in 1996 would be the first one to actually use a GPS radio collar on a wolf (Mech et al. 1998), and I was soon testing these collars in my Minnesota study (Merrill et al. 1998). Still, one would have to drug a wolf to collar it, and that process might forever cause the animal and perhaps its packmates to fear me and my associates. That might disrupt the whole rapport I had with the wolves, which was so valuable. No way was I going to risk anything like that. Thus, each year my approach to studying the Ellesmere wolves remained the same.

In 1989, the wolves failed to use the original den. This development greatly disrupted the research routine. Instead of camping in the usual site of the past three summers, my partner, Mary Maule, and I had to become peripatetic. Rather than sitting near the den each day and observing the wolves' behavior, our task became merely to locate the den. That meant first finding a wolf to follow—essentially starting from scratch, the way I had done during my first summer. While that would be another adventure, it would have been far more productive to be able to continue as in the previous three years.

This time finding the den took much longer. After searching far and wide for two weeks, scanning from hills and mountainsides out over huge expanses of tundra, and watching near the weather station for hours on end for a wolf, we finally scored. On June 27, we spotted the breeding female and followed her in a direction totally different from that of the original den: "We followed Mom as fast as we could although the going was rough. She headed directly NE till she got to the base of Blacktop and then headed N along its base for several miles. By checking my speedometer after, I judged she was traveling about 12 km/hr. Sometimes she would lope for a while."

The area Mom traversed was treacherous for ATVs, so she soon outpaced us and left us with an erroneous idea of the den's direction. This happened several times, and I later suspected that she (the same breeding female that I had observed for three summers) was deliberately trying to mislead us away from the den. "Various kinds of birds use a broken-wing act that leads predators away from their nests, so couldn't wolves do something similar that leads an intruder away from their pups?" I wondered. This behavior was similar to when Yellowstone's 06 Female wolf spent 12 hours decoying a grizzly bear away from her den of pups (McIntyre 2013).

When I later mentioned the possibility of wolves behaving this way to my friend, McIntyre (2019, 2020, 2021, 2022), longtime naturalist and wolf observer in Yellowstone National Park, he suggested that this behavior fit the theory-of-mind idea. This concept suggests that one animal surmises what is in another's mind (Whiten 1991). Did the nursing female I was trying to follow surmise that this was what I was doing? Did that wolf read my mind? After all, that same animal had seen me around her den and pups during the previous three years. Of course I will never know, but I still wonder. The most recent studies in domestic dogs have produced evidence that they behave like the theory of mind predicts (Lonardo et al. 2021; Schüneman et al. 2021). When hearing of my observation, prominent wolf behaviorist Zsofia Viranyi of Austria's Wolf Science Center suggested in a December 8, 2023, email that "this is what cognitive researchers would call tactical deception, although in an ideal case careful experimentation would follow to tackle the exact mechanisms."

During one of my extended den-searching trips (in the wrong direction), a major snowstorm hit overnight while Mary and I were camping on a hillside, and that made it fruitless the next day to visually search the vast, snow-covered expanses for white wolves. A few days later, however, while returning to our food cache at Eureka, we suddenly ran into the whole pack of six wolves, except for Mom, probably back at the den, wherever that was. Several small muskox herds were spread over that area, and the wolves were on the hunt. During the next seven hours, the pack interacted with six herds of adults and calves (37 adults and 11 calves total), having no success with the first five herds.

Then luck changed for the wolves. A herd of four cows and three calves were grouped tightly, and about 200 yards (185 meters) away were nine adults and three calves. In between, stood a calf that was confused about which way to turn. The six wolves headed straight to the calf, grabbed it around the head, and pulled it down.

A few minutes later, the pack also killed a yearling muskox from the herd. They then noticed a single bull "hiding" in a deep gully run up the bank and try to join a herd of eight cows and three calves. The wolves headed straight toward him and tried to attack, but the bull whirled and charged them. Within 30–60 seconds they gave up and the bull ran safely into the herd.

When all this drama ceased and the wolves settled in to devour their quarry, my partner and I settled in nearby to observe. I was certain this find would lead us to the den because certainly some pack members, at least the breeding male, would take part of their spoils to Mom and the pups. All we had to do was wait until the pack finished consuming its catch and follow whichever individuals tore off some prey parts and began beelining it away. When wolves with food are heading to the den, they take the straightest route.

Thus, although Mary and I had been on our way back to our base camp after three days out and had run out of food, we decided to wait out the feeding wolves and then follow them to the den. This was our big chance to find the den and the pups. Even if the pack were to shake us, the wolves' route would at least reveal the den direction.

But it was not so simple. The wolves sure took their sweet time to finish off their two fresh kills (Mech 2011). Gorge and sleep, gorge and sleep, hour after hour. When would they ever finish and leave? I was sure that at any time one or more would break away from the feeding frenzy and head to the den with a load of food. Meanwhile there were Mary and me, both starving and taking turns sleeping so we could be sure to be ready when the wolves did head off to the den.

But the wolves never did. When the carcasses were cleaned, and only well-chewed bones marked the spots where the rest of the muskox calf and yearling had been consumed and was being converted to wolves, the whole pack just arose, strolled off over the hummocks, and headed right back into the area where we had just spent days thoroughly exploring. After all that time and effort, we were not a bit closer to finding the den.

It later turned out that these six wolves, contrary to everything I knew about the behavior of breeding packs, spent much less time attending the den that year than usual. Almost always during summer, the breeding male and other pack members (generally yearlings) do the hunting for the entire pack and fetch food for the female and pups almost daily (Mech et al. 1999).

With this pack in summer 1989, however, the most wolves Rolf Peterson (who later replaced me for several days), Mary, or I ever saw with the four pups and the breeding female (Mom) were Whitey plus a presumed yearling male and a young female. At least that was the case during the ten days that we eventually watched the den and then watched a rendezvous site between July 23 and August 10 (described later in the chapter). We were not certain we ever saw the breeding male there, although we did see an unidentified wolf there for a short time that might have been him.

(During the summer before, one of my Denali packs, the Clearwater pack, behaved the same way, with the breeding male rarely if ever at the den [Mech et al. 1998]. This was the only other time—besides Ellesmere in 2010—I have known that to happen.)

Thus, a month after my arrival in 1989, I had still not found the den, even after spending some helicopter time searching for it. I then took a scheduled break to tend to my other projects, and Rolf Peterson, then director of the Isle Royale wolf and moose research, spelled me for two weeks while Mary stayed on. It finally took Rolf and Mary ten more days to find the den, and then only after some very treacherous ATV travel over the top of imposing Blacktop Ridge. The den turned out to be in an area where I had never been, some 14 miles (22 kilometers) from the original den and in an entirely different watershed.

Rolf and partner, once again following Mom from near the weather station, finally spotted the den through binocs from quite a distance, as Rolf's July 23 notes attest:

> I continued to glass the ridges N of where Mom was last seen, hoping she stuck to a straight course. About 20 minutes later she miraculously moved into view on a low lying whitish ridge just over 3 miles [4.8 km] away to NE of my location. She stopped, nosed around some rocks on the ridgetop, then went west off the ridge and lay down 100 yds away. Soon back on her feet, ascended the ridge again and this time I thought I saw a tiny grey speck trailing her. . . . After more nosing around she went off the ridge to the west again and once again it looked like a tiny grey speck followed her. My hopes this was a den were rapidly rising, but my eyes could still be playing tricks on me. Mom disappeared, but I finely got a firm "fix" on the speck, to see if it moved at all (then it had to be a pup!). Next time I checked it had indeed disappeared and I was sure this was the den.

Rolf and Mary spent the next few days observing at the den, but I then returned and replaced Rolf. By then, the wolves had already moved to a rendezvous site, and it took me a few days to locate that site (August 3), some 6 miles (10 kilometers) from the new den. There, Mary and I observed Mom, Whitey (a 1988 yearling), a yearling male and a yearling female, and four pups. The interactions between Mom and Whitey were especially interesting and will be detailed later in this chapter and also in chapter 10.

Why did Mom, who denned in the original 1986 den during 1987 and 1988, decide in 1989 to den elsewhere, 14 miles (22 kilometers) away? No one can ever know. It would turn out that in later years the wolves also chose other sites, even the year after using the 1986 den. Whatever the explanation, it became clear in 1989 that studying the Eureka wolves was not always going to be logistically as easy as during the first three summers. From a research standpoint, that was a definite disadvantage. It did, however, make the annual expedition much more of an adventure. It also reminded me that having a radio collar on one of the breeding adults would make finding the den each year much easier.

Summer 1990 brought several more surprises. The pack produced only a single pup, perhaps because there was a new breeding female. Whitey, now either a three- or four-year-old and a first-breeder, was the new mother. Mom was peripheral. Not only peripheral, but afraid of her daughter, who even bared her teeth at Mom. The breeding male was the same as in 1988, an adult I had first identified in 1986 by a permanent scar on his left shoulder. The pup was born, not in a den, but in a pit hollowed out of the narrow top of a dirt ridge between two yawning clefts of a deeply eroded riverbank (Mech 1993b). The pit was located 2.5 miles (4 kilometers) from the 1986–1988 rock-cave den, and within a few days after I found the natal pit, Whitey had carried the pup from its precarious and exposed pit den to the cave.

A 1990 mystery that I never solved was whatever happened to all the other 1989 wolves. When I left that summer, there were eight adults and four pups. Now I could only account for Whitey, Mom, and the breeding male. It's not that I spent less time with the wolves that year. Either I and/or a colleague was there from June 20 to August 9. And we traveled as far from the den as 22 miles (35 kilometers). Not only that, but in summer 1991 a pack of seven was seen in the area, and I also spotted extra wolves, namely, those that did not hang around the den. But where were they in 1990?

Another unusual event during 1990 was an exceptionally long move the pack made with the pup when it was only six or seven weeks old, some 22 miles (35 kilometers), to a rendezvous site, which earned the pup the sobriquet "SuperPup." Such a move between the birthing den and a rendezvous site is certainly a record move by a long shot, and, along with the use of such an exposed ground depression as a birthing den, became a major research highlight of the summer (Mech 2022).

That year also brought a personal arctic adventure. My first partner that year was Una Swain, a graduate student, followed later by Luigi Boitani, a fellow member of the IUCN Wolf Specialist Group from Italy. All three of us overlapped during the period when Whitey took SuperPup on the long journey. The little convoy of four wolves and we three biologists headed into an area I had never explored, crossing the Slidre River and far beyond.

A couple of days later, it was time for Luigi to catch a flight out to connect with his long trip back to Rome, so it was imperative that he get back to the Eureka runway in time. All three of us then abandoned the wolves and started back. When we got to the Slidre River, however, we were surprised to find it much swollen. The river runs into the fiord, which is part of the Arctic Ocean, and it turned out the tide was up, thus flooding the river and its flats. Big bummer.

We searched frantically back and forth along the river for a place shallow enough to drive our ATVs across. The timing was such that we could not wait for the tide to subside. We had to find a place to cross—soon. After a few anxious hours, one of us suggested that we locate a narrow spot where the river was also shallow enough for us to carry our ATVs across.

And that is what we sought out to do. After finding such a spot, unleashing all our camping gear off the machines, stripping down, and wading the gear a few meters across the frigid glacier melt off, chest deep, we turned to the ATVs. One by one, we three frozen folks first lifted the heavy three-wheeler high and carried it across, holding it high enough to keep the battery and electrical components dry, and planted it on the shore. Its bulky wheels helped keep it half afloat. Then we repeated that with each of the four-wheelers and climbed out.

Lo and behold, the trusty machines all started!

And I suddenly decided that as long as I was already soaked and cold, I plunged back into the river for a little swim, just to say I had a dip in the Arctic Ocean. The three of us then dried off, packed, and sped back to Eureka, where Luigi made his flight out with a few hours to spare.

Aside from that little interruption, summer 1990 yielded information on some social behavior I had not seen before and had only read about in captive wolves (Goodman and Klinghammer 1985). My wolves started hugging. In hugging, a wolf "puts its front legs around the head and neck of another while each lies on its side chest-to-chest, or on its haunches facing each other, or side-by-side on haunches with one placing front legs around the other's neck" (Mech 2001, 179). I observed pairs of the three adults hugging each other in every combination (table 6.1). Other than in 1990, the only time I saw wolves hugging was the 1990 breeding pair again, once in 1992.

Table 6.1. Description of "hugging" in members of the Ellesmere Island wolf pack (Mech 2001), Eureka, Ellesmere Island, Nunavut, Canada

Date	Description
June 30, 1990	Breeding male and postreproductive female lie on side chest-to-chest and each puts front legs over the other.
July 14, 1990	Postreproductive female hugs breeding daughter from behind three times, sitting, chest-to-back, but rumps were side-by-side.
July 14, 1990	Breeding female and breeding male lie down facing each other, put legs over each other's shoulders, and nuzzle each other.
July 17, 1990	Breeding female and breeding male face each other lying down and female puts legs around male's neck.
July 9, 1992	Breeding female lies with breeding male and places both legs around his neck but he jumps up.

I am still puzzling over this unusual behavior as well as other strange interactions I observed with the same wolves in 1990. The following is from my field notes: "19 July, 0915— Mom headed to the east and Whitey joined her. Whitey did a playbow to Mom and the 2 stood on hind legs and boxed [like red foxes do] for about a minute. Then both continued E with Mom leading." Wolves boxing? Actually, they seemed to be pawing each other while standing, and it was not clear at all what that was all about, possibly some kind of status demonstration like thought to be the case with foxes. I was just bewildered by it.

A week later, I witnessed another set of unusual behaviors, this time between Whitey and her mate when separated:

> 26 July, 1217—Male's . . . behavior is very different—timid, wary, and seems disturbed by us. Our first impression was that this was a strange wolf. At 1251, Whitey appears, arriving from the east-northeast. The male is alert, seems wary at her approach, watching intently. She advances cautiously, slowly, he makes as if to run away, in fact he takes several trotting steps back with tail low, he stops, turns, and whines at her still about 400 feet away. She is moving towards him, he is more or less stationary. They meet—she approaches submissively, ears back, tail between legs, licking up to him. He poses—very upright stance, tail held very high. He moves twice to smell her genitals, she jumps back. This greeting is unlike greetings at den: they hadn't seen each other for >20 hr. Whitey then marked a FLU [flexed-leg urination].

And then I spotted them boxing again, on August 8 at 0113, when suddenly, the male looked intently NW. There I saw another wolf and found it to be Whitey. Clearly neither wolf was sure which the other was. They were about 150 yards (or meters) apart.

Whitey followed me. She went hesitantly to the male, and they sniffed each other and acted like they were mock fighting. This included standing on their hind legs and almost boxing. There was much such sparring, which continued for about a minute and a half in spurts. I was really impressed by how much they stood on their hind legs and by the fact that there were four to six such bouts before they settled down and acted like they knew each other. Although there was no aggression, the amount of sparring seemed unusual.

I have never seen this before, and I wondered whether it was caused by (1) the two mates meeting each other away from the den or rendezvous site or (2) the two mates having been separated for many hours and meeting each other away from the den or rendezvous site. However, I have seen these same two wolves meet each other after 58 hours of separation, but that was at the den, whereupon no such ritual took place.

Was all this unusual social behavior as well as the use of a pit instead of a hole in the ground or a rock cave for a pup-bearing den a result of Whitey's naivete, being the first time she produced pups? I had no way of knowing or gauging this, but it remains a reasonable hypothesis.

Although using a shallow pit rather than a rock cave or hole in the ground as a birthing den seemed unusual, Whitey used that same depression in 1991 when she produced two pups. The pit was atop such a narrow cutbank that when the pups grew old enough to begin climbing out, they could easily have fallen over the edge of the bank. As I watched the pups clamber up the side of the pit almost over the top, I wondered when or if Whitey might take similar stock of the hazardous situation and move them.

At 0517 hrs on June 18, I wrote: "The pups came up to top lip of pit several times and looked precarious there. They could easily tumble over and slide down cliff some 100′. Twice the back edge of the pit crumbled while the pups were in it, but they did not get covered."

Sure enough, at 1207 hrs, Whitey picked up one of the pups and carried it off. Over the rest of the summer, she ended up moving both pups several times, seemingly because she could not find a protective enough spot for them. Finally, on July 8, when the pups were about five or six weeks old, she found what appeared to be an old wolf den at the top of a ridge along Slidre Fiord about 3 miles (5 kilometers) south-southwest of the 1986–1988 den and a bit over a mile west of the pit. There is more in chapter 11 about Whitey's multiple moves and digging new pits.

Whitey kept the pups at the fiord den for the remainder of my 1991 stay. This move was of special interest, for Whitey clearly knew of the other den because she had moved her 1990 pup there and almost certainly was born there herself.

Another interesting development in 1991 concerned Mom's status. For the first several days of my observations, Whitey and her mate, "Blackmask," were the only adults involved. Furthermore, weather-station personnel had indicated they thought Mom had been killed last fall. The word was that in late November one fellow saw Whitey and the 1990 pup (but not the adult male) chase Mom, and Mom was badly wounded around the neck, with her jaw apparently broken. "Her teeth were hanging out to the side," according to the worker who viewed her from 2 meters away. Whitey also had blood on her right hip.

Thus, I had assumed Mom was dead. However, on June 19 at 1054 hrs, Blackmask suddenly charged wildly off the bench he had been lying on and headed NW across a valley and up a long slope. About 200–300 meters farther upslope were two wolves also running NW at top speed, about 50 meters apart. Blackmask soon overtook the last wolf, who immediately submitted as Blackmask drew near. This wolf turned out to be Whitey, and she then followed Blackmask, who continued chasing the third wolf.

All three disappeared over the hill a couple of minutes later. I will never know which the third wolf was, but a few days later I suspected it was Mom. My suspicion arose when that day I got to my observation lookout and a third adult wolf was lying 50 meters from Whitey, Blackmask, and the pups. It appeared reluctant to join the group yet was being allowed fairly close.

I just had to check the identity of that third wolf. I started my four-wheeler, gunned the engine, and waved my arm in a throwing motion that I knew any wolf I had worked with earlier would recognize as a prefeeding signal. The new wolf instantly headed toward me. As I started throwing pieces of fish so I could lure the third wolf closer, both Whitey and Blackmask rushed up and grabbed them. The third wolf hung back about 30 meters, again seeming reluctant to get too close to the others.

There could only be one wolf that pair would tolerate this close, and that was Mom, who had accompanied them for the last few years; and then I recognized her. What I found so interesting was that Mom was only accepted at a distance now, whereas last year she had been an integral part of the pack. What was going on now? How would Mom fare with the pair this year, now being a semioutcast after having been the matriarch of this pack for so long? I was keen to watch for any new developments.

After Whitey returned to her pups, Mom headed that way too and lay about 75 meters from them. A few hours later, at 2256 hrs, Mom headed right over to Blackmask and submitted to him. Whitey then approached Mom and dominated her for several minutes, growling and holding her so she laydown submissively. This was about 15 meters downhill from the pups, and it looked like Mom wanted to get to the pups, but Whitey kept her away. Still, Mom ended up lying a few meters away from Whitey and the pups.

Once again, Whitey chased and pinned Mom with much growling, and Mom disappeared over a hill. Whitey then nursed the pups.

A few minutes later, however, Mom popped up again and very submissively strode to the pups. I noted as follows:

> Whitey stood over Mom very stereotypically for 3–4 minutes holding her mouth above Mom's while Mom pointed her muzzle at Whitey's. Whitey's tail was not raised. Mom was crouched. The two stood in this position for 3–4 minutes. At times Mom was on her back. But one pup crawled over Mom while Whitey was dominating Mom, and Whitey let it. It seemed like Whitey was just checking (i.e., making sure) Mom's mouth was not biting the pup. The pup was trying to nurse from Mom and lay in her groin area for 1–2 minutes.

Mom then headed several meters away from Whitey and the pups and rested.

For the next few days, Mom continued to wheedle her way with Whitey and the pups. She freely greeted and solicited Blackmask (BM) and even sat on him once while he was lying down.

On June 23 at 2007 hrs: "Mom came down the hill confidently toward Whitey, the pups, and BM. Whereas just yesterday she was submissive even when 100′ or more away. Today she did not submit till about 15′ from BM. She then licked up to BM, stood over BM's head for about 40 sec, but BM paid no attention. Then Mom slept about 8′ from BM and 10′ from Whitey and pups."

On June 24 at 2356 hrs we saw Whitey go to the den and nurse the pups. Mom had the pups near her groin when Whitey arrived, but Whitey did not chastise Mom at all, and when Whitey nursed, Mom lay right next her.

The next day, the pups were alone when I got to my observation post. Mom was the first adult to return, followed an hour and a half later by Whitey. Mom ran right up to Whitey and submitted to her. The pups crawled all around and over Whitey, and Mom went back to sleep. A half hour later, Blackmask arrived. Mom and Whitey rushed to meet him, licked up to him, and submitted. Whitey, and then Blackmask and Mom, ran to the pups. Whitey and Mom licked up to the male again, and Whitey nursed the pups. Then all the wolves lay down.

A few minutes later, Whitey tried to carry a pup but was having trouble grasping it. Presumably it was whining. Blackmask awoke and went over to Whitey and the pup and regurgitated. Mom came over and tried to join Whitey and the pups in feeding, but Blackmask pinned her muzzle and growled, preventing Mom from getting much. Mom was now fully accepted but clearly lowest on the totem pole.

With Mom now "officially" part of the pack, Whitey was now freer to accompany Blackmask on hunting expeditions while Mom babysat, which Whitey clearly relished. Before that, Whitey had even once shown disappointment (as it appeared to me and my companion anyway) when Mom headed out hunting with Blackmask while Whitey started out with them but had to return to protect the pups.

Whitey had continually been showing concern about her pups while she centered them in the various pits she dug during their first several weeks of life. Leaving them exposed in shallow pits was far different from leaving them in a hole in the ground or a rock cave, and Whitey seemed to sense that. I never figured out why she didn't move them to the rock cave where she had moved her single pup the year before. Perhaps moving two pups that far would have been too dangerous, for she would have to have left one for quite a long time while carrying the other. With so little cover in that barren area, the pups would be quite exposed (fig. 6.1).

Figure 6.1. With so little cover in the barren High Arctic, wolf pups are quite exposed to any dangers.

Whitey did produce her 1992 litter of three pups in that 1986–1988 den. Then, when they were about five weeks old, she moved them to the 1990 den above the fiord about a week later than in 1991. In 1992, Mom, the babysitter, was gone. Only the breeding pair was present that summer. Still, one or both of them made the first kill of an adult muskox that I had recorded since starting the study in 1986 (see chapter 15).

Another unique finding in 1992 was the high degree of scratching the wolves did. This scratching apparently was related to wolves from another pack frequenting the area and the antagonistic interactions between the packs (Mech 2006), behavior that will be discussed in chapter 21. That year (1992) was also distinguished by (1) the visit of British actor Timothy Dalton (aka James Bond, 1986–1994), who had arrived to star with me in a new TV documentary (https://www.youtube.com/watch?v=dpHZxdzsDj8), and (2) unique information I acquired about trying to extinguish a feeding response one of the wolves had learned.

The latter incident occurred on July 31, 1992, at about 0920 hrs (Mech 2017a). What was so unusual about this situation was the wolf's demeanor and behavior. He was impatient, persistent, and bold as he stalked the food source itself, my bait bucket attached to the ATV's handlebar. Here, paraphrased from my field notes is that description:

> The male wolf arrived, and not wanting to let him nose around camp, I, on my ATV, lured him away. He acted very hungry and eagerly ate even dry dogfood, but he soon started into his predation mode, running after the ATV with tail up and looking at the bait bucket.
>
> The minute I stopped, he came to within 1 meter of the bucket and seemed to try to grab it. Each time, I'd throw some food to him, but each time he would immediately return and act the same way in a very alert mode, ears forward, very anticipatory. I gunned the engine, and the wolf pulled back only 30–60 cm, but he quickly habituated to that. Thus, I was forced to throw him food even though I knew this just rewarded his behavior. The instant I tried to move away, he started his stalking behavior. I gradually moved ~150 meters, but he remained in the stalking mode, and I only warded him off by feeding him some 10–20 times. This continued for ~20 more min.
>
> I tried various means to change the wolf's behavior without just throwing him the food bucket, which held the seal meat. I switched the bucket from the left side of the ATV, where the wolf was, to the right

> side, but he just switched his stalk to the right side. I held the bucket on my lap, but he still seemed about to grab it. Once I took a handkerchief out of my coat pocket and the wolf started eying my pocket.
>
> The way I finally broke the wolf of his stalking was to put dirt clumps into the bucket, breaking small pieces off like food bits and showing each to him like food and then tossing it to him. He went after the piece each time, and a few times he grabbed and bit them and spit them out. I tried this about 50 times, and each time he would go after the clump. After 5–10 min of this, the wolf started to leave. When about 4 meters away he stopped and resumed his stalk. I resumed throwing him only dirt clumps. After ~20 more, he started away again but returned once more. I threw more clumps, and he checked each.
>
> Finally, the wolf trotted off ~200 meters. As I drove off, he came running back and continued the same stalking behavior. I continued throwing him dirt clumps, and he went after about 20 more. After ~5 min, he left, and I let him get ~200 meters more away. I then left in the opposite direction at about 1030 hr, and the wolf did not follow. (Mech 2017a, 24)

All this plus a fatal attack by the pack on a strange wolf during 1992, detailed later, made that summer fruitful and exciting. In fact, this whole new phase of these wolves' lifestyle, from 1989 through 1992, was both exciting and fruitful for my studies. It forced me out of my 1986–1988 routine and into exploring new areas of the island and positioning myself to observe a much wider variety of behaviors.

:: 7 ::

A Whole New Phase

The next year (1993) was unusual in that when I arrived, the wolves were not denning at all; apparently they had produced no pups or else the pups had already perished. A thorough search of all the areas where the wolves had denned so far, except for the distant 1989 den, showed no sign of new usage, and weather-station personnel had been seeing five adult wolves traveling together and had not noticed nipples on any of them. Visible nipples would mean the animal had been nursing pups recently. Thus, I reasoned that the pack of five could be the breeding pair and last year's pups. If so, that could also mean that the pack had no new pups because usually when they have pups at a den, the adults do not travel together until late summer. At least one of them, usually the mother, stays with the pups while the others hunt.

This new development, after a six-year routine of camping near the den and observing the wolves daily, posed a major challenge: how to structure the study when the wolves' daily locations were no longer predictable.

Find the wolves and follow them was the obvious solution, but how to do that was the greater problem. Without radio collars on any of the wolves, trying to find them when they are not denning and thus are nomadic over an area of perhaps 400 miles^2 (1,000 kilometers^2) or more could be quite problematic. The question even occurred to me as to whether it would be more worthwhile to just skip that summer and return to working with my Minnesota and/or Denali study, which assistants and students were conducting while I was gone.

I still had to check the 1989 den and rendezvous site, which were much harder to reach than the others. It turned out that I didn't need to. Within three days, the pack of five visited the weather-station area while my two assistants and I were still getting organized and figuring out what to do. The wolves, being white, were easy to spot from a distance over the wide-open terrain. They responded to me the same way they had a year earlier, so I was able to identify the two adults as Whitey and her mate, and by the way the other three acted, I could assume they were the three 1992 pups, now yearlings. And, true, although I confirmed Whitey's identification marks—a notch in her outer left ear and a missing top half of her right canine—I could not see any active nipples.

Based on my attempts to follow wolves on their daily travels from the den during the previous summers, I knew that spending this summer trying to find and follow them would be challenging. At times I had been able to tail them for long distances over terrain that was easy for ATVs, such as along river bottoms or over fairly smooth, flat areas. But when the animals easily traversed steep canyons or nimbly strode across the many flats carved by hummocks of various sizes and shapes that the ATVs had to very slowly crawl over, I quickly lost them.

Still, I had to try. Doing so would greatly add to my knowledge of the extent of the pack's home range and might sometime lead to my seeing the animals hunt. Watching hunts was always informative and exciting. By this time, I had watched four successful hunts, but one was the one described earlier that I had anticipated after spotting the muskox herd from the den area, so that wasn't really while I was traveling with the wolves. With two other hunts, I had been searching for the wolves and found them when they were about to encounter prey. Thus, I had only watched one successful hunt that had taken place while I was following the wolves and thus able to observe the entire sequence the way it unfolds most of the time; that is, from before either the wolves or the prey had detected the other.

Each hunt was different so the potential for new information was high, whether the wolves were hunting for muskoxen or arctic hares, or possibly even caribou. From how far away did the wolves detect the prey? How did the hunt start? Which wolf started it? What did each wolf do? How did the hunt proceed? When did the prey detect the wolves? What did the prey do? How long did the hunt last? Was it successful? What did the wolves do after? What did the prey do after? We learned, for example, that when a pack succeeded at a hunt and gorged, the parents quickly returned to the den to feed the pups, while the remaining members tended to spend more time around the kill. There was always so much to watch and record, but it would only be after many such observations that any generalization could be reached.

Besides the hunting behavior, there were always the travels themselves. Specifically, where did the wolves go? How far and fast? How did they interact? Which wolf led? Where and when did they scent-mark? Howl? Which wolves were involved? How did the others act? When did they rest? Sleep? How long? Which wolf initiated any of these actions? After spending so much of my career in an aircraft watching wolves of unidentifiable age and sex far below traveling and hunting, to be able to join those creatures on the ground and travel with them while knowing their age and sex was an incredible opportunity.

Thus, I spent several weeks in 1993 intermittently following the five wolves until during each bout I lost them either because topography thwarted me or because they left while I slept. Then it often took days before I would find them again or they would reappear near my camp. Once one of the yearlings, which I had dubbed "Explorer," came too near my camp, this time when my future wife, Laurie, had joined me for a couple of weeks. My field notes told the story:

> She [Explorer] came to within 6 feet of us and was interested in our food. I had to scare her away by jumping up and shouting "No" to her or she would have come closer. As it was, she circled camp sniffing the 4-wheelers and tents, and then rolled in our sleeping bags which we had out airing. When she tried to run off with our double sleeping bag, about 50′ away from us, I had to run at her with 2 pot covers, clanging them like cymbals. She took the bags a few meters and dropped them. Each time she came back to them, I started half-crouched, toward her, just lightly rubbing the covers together as threat, saving the complete charging and clanging as my ultimate weapon if she did not back off. (I suddenly understood the reason and functioning of the low intensity threat such as a growl and snarl before the full lunge—it saves the latter for when really needed and minimizes habituation to the ultimate attack.) The low intensity threat worked—Explorer left the bags alone and headed back to her pack.

Following the pack in 1993 was fruitful, especially for timing pack travel rate (Mech 1994a), observing den digging despite lack of pups (Mech et al. 1996), and watching the pack hunting arctic hares. During that summer, I saw the five wolves catch many hares, and it seemed that during the period I was there (July 1 to August 6), hares were the main prey of the pack (Mech, Smith, and MacNulty 2015).

It almost seemed as if the wolves were not interested in muskoxen right then. In fact, one day (July 14) at 1407 hrs, the pack encountered two bull muskoxen and swirled around them for about 30 seconds. One wolf got behind a bull and bit him in the rump, but all the wolves then continued on. On the other hand, a few days later about midnight, the breeding male brought Whitey what appeared to be a muskox calf leg, which she fed on for about 10 minutes and then cached. Had the pack nailed a calf during one of the several periods when I had been unable to stay with them?

During summer 1994, my usual study routine around the den returned to normal, or at least as normal as literally living with a pack of wolves can be considered normal. Pack composition that summer included the same breeding pair with two of the 1993 immature members (one of each sex), plus a single pup at the original den. It was a good summer for obtaining data about the interactions between the immature pack members and the rest of the pack, data that were later used in publications about provisioning food to the breeding female and her pup (Mech, Wolf, and Packard 1999) as well as the relative roles of the breeders as pack leaders (Mech 2000b). Intrusion on the den by a muskox proved to be a catalyst that showed which pack members demonstrated their defensive mettle and how (chapter 18).

I also observed the 2-year-old female introducing the 28–33-day-old pup to live food in the form of an arctic-hare leveret (Mech 2014). The wolf had lain the leveret down several times and could have killed it with one bite but did not. The pup tried to eat the young hare as it was flopping around, and finally the 2-year-old joined in and they both consumed it. Although such behavior is well known to housecat lovers who have watched mother cats bring live mice to their kittens, with wolves no one had observed it, or at least reported on it. This was certainly a moment of learning, and to watch this was to become convinced that it was also a teaching moment.

Summer 1995 proved to be a mystery. I checked all the dens, including the distant 1989 den and rendezvous site (by helicopter). The original den showed a bit of fresh digging but no sign of regular use, which by that date (July 4) should have shown weeks of use were it active. What I thought was pup fur in the surrounding vegetation may have evinced pup demise. A pup or pups could have perished soon after birth and been eaten by the adults. This theory that the den did hold pups at some time this year was bolstered by a visit to the den area on July 5 individually by both members of the breeding pair, and by Whitey's entering, digging, and sleeping around the den, along with her 3-year-old male offspring, for at least 13 hours (Mech et al. 1996). No nursing nipples were evident on Whitey, however.

I never was able to determine whether pups were born and lost in 1995, whether Whitey had resorbed her fetuses, whether she never was bred, or whether she was pseudopregnant. Pseudopregnancy is a condition in which an animal seems to be pregnant but holds no fetuses; pseudopregnant individuals behave like they are pregnant, for example, digging dens. Nevertheless, by considering these den-digging observations along with those from other studies, these findings led to a hypothesis that den digging did not require the production of pups but rather might be merely a function of seasonal prolactin secretion (Mech et al. 1996).

Summer 1996 brought more like the original den-watching routine again, with the same breeding pair as in 1994 and no other adults, occupying the original den, which a colleague and I observed from June 26 through August 1. As wild wolves go, these two were getting old. Whitey was born in 1987, and her mate in 1985 or earlier (Mech 1995b, 1997a). They produced two male pups in 1996, and the situation was such that I could study the development and activity of the pups around the den from the age of about 3 to 7 weeks, of which there had been little information.

My summer 1996 study, however, was not entirely uninterrupted. A few weeks after my arrival, I learned that some human wolf-watchers had found their way to Eureka and were camped in the area. Not surprisingly, two European adventurers had got wind of how easy Ellesmere wolves were to photograph and had made their way to the island to test that themselves.

Realizing that the addition of two more humans around the den with their own agenda could disturb the wolves' normal routine, I made a point of addressing the visitors, explaining that I had the only government permit for observing the wolves around the den, and seeking their cooperation. The men had been able to photograph some of the wolves when they visited old muskox carcasses several kilometers away from the den, so I hoped the visitors would be satisfied with that.

A couple of weeks later, on July 25, partner Jeff Turner and I were saddled on our ATVs east of the den and had just observed Whitey and the pups dashing 100 meters to the breeding male and gulping up his regurgitant in mid-morning. As my notes read,

> All of a sudden, about 1040, the wolves began acting very upset. Whitey was just W of the den entrance but O.S. [out of sight] and the pups were visible behind her. They seemed very afraid of heading toward the den, although they seemed to want to, like she was calling them. The male was walking alertly on the N slope and looking into the creek valley just NW of the den, and he crossed back and forth, and went to top of N slope and howled. Meanwhile, the pups finally went up to the den and Whitey led them S to S end of den rocks and then SE away from the den.

The only thing that very strange behavior could mean was that the upset pack was responding to a strange human in the valley below the den. The only possible person who could have been there was one of the visitors. Thus, I headed over to the west side of the den ridge and tried to see down into the creek bottom, but the steep banks prevented that. Meanwhile, the male wolf kept circling that spot and looking down into it. Then I spotted fresh human tracks in a snowbank where someone had come up to the ridge and peered over.

I started yelling down to the creek valley for the intruder to come out, and Whitey came over to my side and peered into the valley with me. Although I have no notes about the following, my recollection is that when Whitey joined me, she also began bark-howling as I hollered (Mech and Janssens 2022). After a few minutes of this, the photographer showed his face. I ordered him out of there, and he finally emerged and struggled up the slope. I explained again that the fellow had no permit to be around the den and that he was clearly disturbing the wolves and thus the study.

The guy finally left, and after I reported the intrusion to the officer-in-charge (OIC) of the weather station, the OIC made certain the photographer would obey. Although no wildlife or police authority was stationed at Eureka, the only government authority of any type there, except sometimes the military, is the OIC.

In addition to all the new wolf information I had gleaned that summer, I managed to make some rare observations about caching and the wolves' communication abilities as well as the breeding male's endurance when hunting hares (Mech 1997a). I also had accumulated enough data about the comings and goings of the adult wolves to detect certain patterns (Mech and Merrill

1998; discussed in chapter 17). One time all the adults were gone for more than 24 hours. I'll say more about these topics later. Suffice it to say here that the 1996 study yielded much good information.

But what was happening to the Eureka wolves? Why, after litter sizes of 4 to 6 pups during 1986 through 1989, did this pack now experience such a low pup production: a single pup in 1990, 2 in 1991, 3 in 1992, none in 1993, 2 in 1994, none in 1995, and 2 in 1996, an average of about 1.5 per year. The most obvious difference was a different breeding female. Whitey took over from Mom as breeder in 1990. With her being a first-breeder, a decreased number of offspring her first year would not be surprising, but that trend continued for seven years.

The problem could also have been inbreeding depression. The new breeding pair seems to have been closely related, for in 1986, Whitey's mate since 1990, a 1–3-year-old in 1986, lived in the same pack with Mom (Mech 1995a). Thus, that male probably was one of Mom's offspring or at least was closely related to her, and in 1987, Whitey was presumably also born of Mom (discussed earlier). Wolves try to avoid inbreeding (Deborah Smith et al. 1997; Mech and McIntyre 2023), and inbreeding depression can result in reduced litter sizes (Laikre and Ryman 1991).

Later studies of DNA from wolf scats, hair, and other specimens did indicate that the Ellesmere wolf population had a lower genetic diversity than mainland wolves farther south and that of other arctic islands (Carmichael et al. 2008; Frévol 2019; Frévol et al. 2023), an indication of being more inbred. Being at the extreme edge of the wolf's geographic distribution, Ellesmere wolves, especially those around Eureka toward the north end of the island, would have much less chance of encountering dispersing wolves from the main wolf distribution thousands of kilometers south.

In fact, it turns out that Ellesmere wolves are genetically distinct from most other arctic wolves and have been called by geneticists "a novel and highly distinct Polar wolf population" distinct from other High Arctic wolves (Sinding et al. 2018) Although I didn't know it then—actually no one did—they also have especially interesting genes that adapt them to living in such a harsh environment as the High Arctic, genes that affect not just their white pigmentation, but also their vision, metabolism, and immunity (Schweizer, Robinson, et al. 2016; Schweizer, vonHoldt, et al. 2016).

In any case, I was anxious to see what I would find in 1997. Would all four of the 1996 wolves have survived? Where would they be denning? How many pups would they have this year?

What a disappointment I was in for. When I arrived on July 1, I learned right away that none of the Eureka Weather Station or military folks had seen the wolf pack since April, and even then they had seen only three members. Furthermore, Whitey was not among them. I soon found that none of the usual dens or rendezvous sites used anytime during the last ten years showed any sign of recent use. What had happened?

The weather-station personnel did mention that in early winter the pack of four had frequented the area often and the folks had been feeding them. The pack had also begun headquartering in a bed of straw some 200 meters from the station. Four adult wolves would require in total a minimum of about 28 pounds (13 kilograms) of food per day just to survive (Peterson and Ciucci 2003). If the station folks fed the pack that amount daily, it would have soon depleted the station's limited meat supply, so certainly anything they fed the wolves would have been just token amounts.

However, the fact that the wolves had headquartered for so long near the station instead of roaming nomadically in search of prey indicated either that vulnerable prey were scarce or that the wolves were incapable of finding and/or catching enough of them. Whatever the case, there was no way of finding out. It was apparent that 1997 was to be a bust for me. If the pack had not been around in two months, there was little chance I could even find them to follow them around. Thus I decided to cut my stay short that year.

Before leaving right after learning that the wolves were nowhere to be found, however, I decided to check the immediate area around the weather station to at least see where the wolves had stationed themselves. Maybe I could find some scats there to examine or check out the straw pile the wolves had lain in.

I did find beds where the wolves had lain in the straw. However, I also discovered something far more grim. From my July 5 notes: "Found remains of a wolf—much fur, a few ribs and small pieces of bone. One long bone with fatty marrow was largest piece. Best hypothesis was that this was remains of Whitey, and she had died, and the others [her mate and pups] ate her."

I had often found wolves in Minnesota that had been eaten by other wolves, presumably their packmates in many cases. Most often the victims were underweight pups in autumn that had probably starved to death. Fall is an annual pinch period when pups' food requirements have increased because of their greater size, yet young prey have already been eaten or grown and become less vulnerable.

In this case, however, it appeared that Whitey had not starved, given that there was still fat in one of her leg bones. She must have died from something else, and then, when her body became available, become food for the rest of her pack. A fur mat and various cracked bones lay about 10 meters from where her pack had bedded, but other bones were scattered as far as 150 meters away. I scraped some dried muscle tissue from a bone to have the DNA checked later for the animal's gender to make sure this really was Whitey. If the critter was a male it could not have been Whitey, but the DNA did later show this was a female.

Although there was no way of knowing how Whitey had died, there was a theory. One of the weather-station people had taken a special interest in the wolves and noted that Whitey had suddenly disappeared. Here, again from my July 5 notes, were my speculations, without estimating any particular date for this occurrence: "Pete mentioned that the day Whitey disappeared he had overheard a couple of military people talking excitedly about some experience they had just had on their snowmobiles up at the airstrip. When they saw Pete, they suddenly clammed up as though they didn't want Pete to hear them. That same day, the 'Old Man' wolf was also very leery of coming up to the weather station where usually fed. Pete wondered whether the 2 military people had chased the wolves and injured Whitey."

Thus, 1997 yielded little wolf information other than the lack of denning and the loss of the breeding female. The OIC of the Department of Defense, who had been headquartered there for a few months, told me that the last time they had seen any wolves around was in late May, when they did see the adult male and what they assumed to be two yearlings.

Neither did 1998 produce any more behavioral data about interactions between the wolves and their pups. In fact, the Eureka wolves did not den in the area that year either. I was able to find a pair of wolves, however, and one was familiar to me. She was one of the 1992 pups, which I had known as the yearling Explorer in 1993. I was able to identify her in 1998 by her comfort and familiarity with me and my feeding routine as well as by her being the only female born in the area since 1991. A couple of dark streaks on her white back confirmed her identity.

Of interest was the fact that her mate must have been a disperser from a pack much farther away. This wolf's fear and reluctance to come any closer than about 200 meters suggested he had originated from an area where wolves had been routinely harassed in some way by humans. He was the only wolf of very many I encountered around Eureka that behaved so fearfully.

Conceivably, other wolves inhabit Ellesmere that never interacted with humans yet are still afraid of them but just haven't been reported. However, even wolves on the other of the Queen Elizabeth Islands were much tamer. Because wolves can disperse more than 600 miles (1,000 kilometers) from their natal homes (Wabakken et al. 2007), Explorer's mate could have been from near any of several military bases or Inuit villages within that distance or from even farther away.

Along with this interesting outsider wolf in 1998 and the recurring lack of denning came two other types of new information. First, my colleague and I were able to observe this pair of wolves encounter, attack, kill, and eat an adult muskox, including some aspects of caching behavior that had not been reported before (Mech and Adams 1999) but will be discussed later. Second, we had found the remains of nine muskoxen that appeared to have starved, and we observed no muskox calves or arctic hare leverets. It became apparent that some kind of unusual environmental conditions had prevailed the previous year.

Although my main interest during my annual Ellesmere visits had always been studying the social and predatory behavior of the wolves, I could not help but be interested in the ecological aspects of this High Arctic ecosystem as well. Most of my time each summer had been spent finding the den and then frequenting the area and watching the wolves there or following them around. Thus, about all the time I had to investigate the prey to any extent was used to record incidental observations of the arctic hares, muskoxen, and caribou. Nevertheless, it was clear in 1998 that something was amiss. Why so little or no prey reproductive success that year? I vowed to try to figure out what the problem was (Mech 2000a).

I was especially eager for my next trip to Ellesmere to see what all was happening, not only with the wolves but also with the muskoxen and hares. But that was not to be. I did make it all the way to Resolute Bay, the last commercial leg of the trip where, as usual, I then was scheduled to catch a government flight to Eureka. However, my lower back rebelled on the flight to Resolute Bay. An old back injury caught up with me and laid me up in Resolute for several days until I could return to Minnesota (transferring between flights while lying on baggage carts). There I was laid up for two months.

Thus, summer 1999 became the only summer when I had no personal information about the wolves at Eureka. I did learn via phone calls with weather-station folks that they had seen Explorer and her mate several times in April and May. However, contrary to other years, neither military nor weather personnel observed wolf pups or other evidence of wolf reproduction that year either (Mech 2004).

It turns out that 1999 gained importance related to the Ellesmere study in another way altogether different from the fieldwork, however. That is when my article proposing a radical switch in the behavioral terminology used to discuss wolves was published, and it was working with the Ellesmere wolves that had inspired it (Mech 1999).

The issue involved the application of the term *alpha* to the breeding pair of wolves in a standard pack. *Alpha* as used in behavioral studies applies to individuals which via their aggressiveness dominate other members of a group. Behaviorist Rudolf Schenkel applied the term to wolves in his study of a group of unrelated wolves that he had gathered together in a captive colony (Schenkel 1947). Like many such groups, the wolves fought and established a dominance hierarchy, or "pecking order." The wolves formed both a male and female order, and Schenkel called the most aggressive ones at the top of the order the "alpha male" and "alpha female."

I referred to that work in my 1970 book, *The Wolf: The Ecology and Behavior of an Endangered Species*, which was a bestseller and remained in print until 2022. Thus, the alpha term came to be used widely among wolf researchers for the top-ranking members of the wolf packs they studied. However, the more I associated so closely with the Ellesmere wolves each summer, the more I wondered why I was calling the parents of my wolf family the "alpha male" and "alpha female." I had long known that those animals had gained their status in the pack merely by mating and producing offspring, not by fighting or battling their way to the top of their family.

Thus, the term *alpha*, implying a status gained by force, was a misnomer. My alphas were merely parents. With wolves, being fierce predators that already had a bad rap with much of the public, adding *alpha* only tends to exaggerate their aggressiveness. Therefore, I decided to discuss the derivation of the alpha term and propose that it be abandoned at least for standard family wolf packs (Mech 1999; Packard 2003; Boyd et al. 2023).

For more complex packs, such as some of those in Yellowstone, which comprise only 10–15% of wolf packs overall (Mech 2024), the term might be more valid. Those packs include a matriarchal female with a daughter or two that is also breeding (Stahler et al. 2013), so distinguishing the matriarch as the alpha female from her breeding daughters can be useful (Mech 1999). Stahler et al. (2020) agreed with this point, as did Ausband (quoted in Pappas 2023). In this case, it only implies that the matriarch is dominant to her daughters, not that she fought to gain her position.

:: 8 ::

Beyond Wolf Behavior

Summer 2000 was really not much better for my Ellesmere study than 1999, except that at least I made it to the island. However, again none of the known dens was occupied, and I only located a pair of wolves that were not denning. I was quite sure the female was Explorer based on her reaction to me, and I got a good look at her abdomen while she sat a meter or so away and stared at me. Her lack of any distended nipples definitely showed she had not nursed that summer. Her mate was not as flighty as two years before but still was pretty stand-offish. In addition, it became clear that if this pair had produced pups in 1999, the year when I failed to make it all the way to Eureka, none had survived. If they had, there would have been more animals with this pair.

Again, I was faced with the kind of dilemma that plagued me during the previous years when the local wolves failed to den, or at least failed to den any place where I could easily reach and study them: (1) how long to remain on the island once I got all the way there, and (2) for how many summers should I continue the study? Not having the wider perspective of where the Eureka wolves lived year-round, it was hard to know what to think about whether there might be wolves other than Explorer's pair denning some place farther away that I could find and study.

My knowledge and experience on Ellesmere extended over an area of only about twice the spread of Chicago, some 22 by 22 miles or 485 miles^2 (35 by 35 kilometers, or about 1,225 kilometers^2). At various times, I had followed wolves as far away as 22 miles (35 kilometers) from the den just in a single day. Other times, I had taken helicopter flights a bit farther out from this area.

But there certainly was much more area of just the Fosheim Peninsula that I did not know.

In the years when the Eureka wolves were not occupying dens when I arrived, what was the reason? I almost always arrived in late June or early July, when pups were a few weeks old. If there were no pups in any given year yet the adult pair was still around, why was that the case? There were many possibilities, but I had no way of getting information that would allow me to distinguish among them. Did the pair not copulate, and if not, why not? Did they copulate but the female did not conceive, and if not, why not? Did the female conceive but resorb her fetuses, like any canid can do if they fail to procure enough food? Or were pups born but died before I arrived?

During summers when the Eureka wolves failed to reproduce, they traveled nomadically over a large area (as discussed earlier for 1993). That year I was able to find them and follow them at least a few times, but I often lost them for days on end. If I could find them more easily, at least my time there would be better spent. Thus, year after year, I was haunted by the temptation to try to radio-collar a wolf or two so that I could readily find them and keep track of them. Or, even better, I was tempted to use the new GPS radio collars, which would also transmit location data to my computer year-round.

The questions about wolf denning were important to try to answer for several reasons, not just for the sake of the research. First, how long should I try to continue this project? The logistics were horrendous. Flights to and from Ellesmere were expensive and time-consuming. Generally, the itinerary required flying from Minnesota to Edmonton, Alberta; overnighting there, then heading to Yellowknife for another layover; on to Resolute Bay the next day if possible; and finally hanging out in Resolute for one or more days before catching a flight to Ellesmere.

The last two legs were often problematic because the town of Resolute Bay (population 198) lies on the south shore of Cornwallis Island, where fog often enveloped the airstrip, and sometimes came and went. Thus, a flight from Yellowknife might be cleared to proceed but a couple of hours later when it approached to land, the island would be fogged in, so the plane would have to turn back. During the first few years of the study, flights failing to land in Resolute Bay would then skip the town and head south 2,125 miles (3,400 kilometers) to Montreal, Quebec. I once had to do that, staying overnight in Montreal and then heading back for four and a half hours to Resolute Bay the next day. Even then, fog threatened to prevent landing again but cleared just in time. Otherwise, the next stop would have been all the way back in Yellowknife.

Once in Resolute, moreover, logistical problems grew more complicated. Flights from there to Eureka are unscheduled charters. Most often they are those chartered by PCSP, the Canadian government agency that subsidizes arctic research. That is a wonderful agency without which most research in the Canadian Arctic could not be done. It provides a headquarters with a bunkhouse in Resolute Bay, rental ATVs and other equipment, as well as flights to several areas in the general region. Proposals for PCSP summer flights to Eureka, however, had to be made the previous October. (Because I had no knowledge of this agency when I first started the project with *National Geographic*, my tradition generally was to stay with one of the few private hotels or tourist homes in the Inuit village there rather than with PCSP, even though I depended on their flights.)

Besides PCSP flights to Eureka, there were miscellaneous other chartered flights there. Some brought food and supplies or construction equipment to Eureka. Others took tourists to Ellesmere's national park but stopped at Eureka to refuel. Some were individual adventurers headed to the North Pole. All these types of flights funneled through Resolute and were serviced by only one or two private charter outfits, and almost the only type of aircraft used were Twin Otters, which usually carried up to 10 or 12 passengers plus considerable cargo.

From Resolute Bay to Eureka generally cost many thousands of dollars per flight. Thus, it was common for whoever was heading to Eureka to pool with others from other projects, who might be arriving in Resolute within a few days. The wait in Resolute Bay for other passengers could take days. And then both Resolute and Eureka weather had to factor into the flight schedule as well. There aren't many alternate airstrips to land at if weather prevents a landing at Eureka. The same problems beset return flights as well.

These logistical and time-consuming issues meant that my decisions to continue research on Ellesmere in any given summer could not be made lightly. If I could know there was a good chance that wolves would not be denning and thus could not be predictably observed around a den, then it would be imprudent to consider the trip, at least after having experienced it several times. The only problem was that because flight arrangements from Resolute Bay to Eureka and back had to be made the previous October, there was no way of judging whether wolves would be denning. Therefore, I applied every fall to make the trip the next summer.

Once at Eureka, the main consideration was to determine whether wolves were attending any of the dens, or at the very least to learn whether any were frequenting the area. If not, there would be no sense in remaining there. I had all those other projects to attend to at various times, always the Minnesota work, but from 1986 to 1995, the Denali project, and from 1995 onward, the Yellowstone reintroduction and research. As it was, the time I was spending on the island had been diminishing down to one to two weeks, mainly because of the lack of denning wolves in 1993, 1995, 1997, 1998, and 2000, and probably not in 1999.

On the other hand, the Ellesmere opportunity was so great, and I had such a rich background of information about the wolves there (and a strong personal attachment to the wolves and the work), that each year's findings grew more valuable. And there was always my intense curiosity about what was happening in my High Arctic study area each summer. I could not help but become increasingly interested in the wolves' hunting (fig. 8.1), traveling, scent-marking, territoriality, and all other aspects of their ecology, just like in my other study areas.

And now with a series of years during which no surviving pups were produced, that broader interest was becoming my main focus. The fact that during 1998 both arctic hares and muskoxen had failed to produce surviving

Figure 8.1. I could not help but become increasingly interested in the wolves' hunting and feeding behavior. Here the wolf pack is pulling down a muskox calf.

offspring really intrigued me. Thus, I decided that I needed some kind of regular assessment of hare and muskox numbers each year. There was no way to get an actual count of these creatures over the entire study area, for that would have required regular flights each year. The smallest and least expensive aircraft available in the region were the Twin Otters or helicopters, both of which were too expensive for such a task.

Rather, the prey assessments would have to be some kind of simple indices that could be done quickly and easily via the ATVs. Of course, I had general notes each year about the muskoxen and hares, and I had already begun some actual counts in 1998. However, I now realized I should repeat them each year.

For the hares, I decided to annually count the number I saw along an old dusty roadway that ran from the Eureka Weather Station about 5.5 miles (9 kilometers) along the north shore of Slidre Fiord and up to an international atmospheric research lab at the top of a mountain, the Polar Environment Atmospheric Research Lab (PEARL), or Astrolab. The hares' white coats could be seen for a long distance from the roadway, no matter whether along the shore, on the side of the hills above the roadway, or from the mountain.

Muskoxen are nomadic about the area, so no such transect approach like counting along a a standard set of routes over several years would have worked for them. However, there was a large expanse of about 100 miles^2 (260 kilometers^2) of plains that muskoxen often frequented, some part of which could be surveyed with binoculars or spotting scope from the side of a mountain known as Blacktop Ridge. It was always such a pleasure to sit up there each year and search for the dark, shaggy beasts over such a large area, when we sometimes counted over 100 of them.

I continued both the hare and muskox index counts every year throughout the rest of the study and eventually found that what I was learning about the hare population turned out to be critical to the local pup production and/or survival each year (Mech 2005, 2007a; detailed later). That made it worthwhile to continue annual trips to Ellesmere even if in any given year the wolves did not den or even if no wolves were even around. And, in fact, the latter turned out to be the case for a few summers.

In 2001, I saw not one wolf during a 13-day period there and tracks of only a single wolf, and none of the known dens was occupied. The only report I had from weather-station personnel was that, on June 9, 2001, three wolves had harassed a polar bear near the weather station. About that interaction, the station's June 11, 2001, email read: "they appeared to be curious rather than hostile . . . nose to nose they checked each other out."

What I did find in 2001, however, were the remains of 17 muskox carcasses, far more than I had ever found before (table 8.1). Of great interest, besides the unusually large number of carcasses found, was the fact that most of them showed no fat in their bone marrow, similar to nine carcasses I had discovered in 1998. Bone-marrow fat is an ungulate's last fat store (Mech and DelGuidice 1985), and it only comprises 2–3% of total body fat (McCullough and Ullrey 1983). When it is gone, the animal perishes from malnutrition or is extremely vulnerable to predation or other direct mortality factors.

Table 8.1. Characteristics of fresh muskox remains found during summer 2001 in the Eureka study area, Eureka, Ellesmere Island, Nunavut, Canada

Sex	Estimated age (years)	% marrow fat	Estimated season died	Eaten by wolves?
	Adult	0	Winter	Yes
	Adult		Spring	No
	Calf	100	Fall/Winter	Yes
M	10	0	Spring	No
F	12–15	0	Winter	No
	Adult	0	Winter	Partly
	Adult	0	Winter/Spring	Yes
	Calf	0	Winter	Partly
	Calf		Winter	Yes
M	Adult	0	Spring	No
	Calf	0	Winter	Partly
	Calf	0	Winter	Partly
F	10	0	Winter	Partly
	Calf	0	Winter	Partly
	Calf	0	Winter	Partly
	1	100	Winter	Yes
F	7–8	0	Winter	No

Note: Percentage marrow fat was estimated visually, with the bones being either hollow (0%) or the marrow appearing fully fat (100%). Partly eaten remains were assumed to be those of animals that had died and were eaten as carrion by wolves. Those indicated as eaten by wolves were probably also killed by wolves.

What happened? Why couldn't so many muskoxen make it through the winter? And was this discovery in any way related to my 1998 findings that neither muskoxen nor arctic hares had successfully reproduced? That, too, could well be another possible reflection of the lack of adequate nutrition. This would take some pondering and further investigation beyond just my fieldwork on Ellesmere. Rain-on-snow or melt-freeze cycles can lock up forage and contribute to overwinter muskox mortality exceeding 90% (Miller and Barry 2009). However, the Eureka area is generally sheltered from some of this most severe icing.

Again during a short visit in 2002, I found none of the dens active, no wolves, and very few tracks in the study area even though weather-station and military personnel had reported a pack of four wolves being seen in the area several times in May and July.

Finally, in 2003, a pack of three wolves was again visiting the Eureka area, including what appeared to be a bonded pair based on their double scent-marking. However, it was clear this pair had produced no surviving pups in 2003, and again none of the dens had been used. Two of the three wolves showed a great deal of familiarity with my feeding routine, so I knew they were those I had worked with several years earlier, perhaps in 1998. However, the male of the 1998 pair was very fearful whereas the male this year was exceedingly tame, so this animal might have been one of the 1996 pups. The weather-station crew also knew this wolf, the largest I had seen on Ellesmere, and referred to him as "Brutus."

Brutus's bold behavior was so distinctive that he became the star of the next few years of my behavioral observations. With Brutus on the job, the Eureka wolf population began to thrive again. Brutus and I even sort of became buddies, although during one of the interactions between us, I wasn't sure he was so keen on that. During our first interaction, in 2003, Brutus had come up and sniffed one of my gloves as I grasped the handlebar of my ATV. That was fine. It was a year later that Brutus seemed to get a bit too friendly.

In 2004, Dean Cluff joined me on Ellesmere. I had known Dean for several years through annual International Wolf Center trips to observe wolves in the Aylmer Lake area northeast of Yellowknife, Dean's wolf-study area for the Government of the Northwest Territories. In 2004, Dean and I found that the usual Ellesmere dens were unoccupied but did encounter a nursing female.

While trying to find the female's den, we two were hot on her trail following and tracked her for several kilometers on our ATVs to an area we believed must be close to the den. Then we had to start tracking on foot up a rock-strewn creek bed, having left our ATVs on a hillside above a canyon I thought was too steep to navigate on the machines. Suddenly we spotted a wolf up ahead.

According to my July 10 field notes:

> As we watched, the 2nd wolf popped back and forth over the slope into view, and finally we spotted a pup coming over the slope, and eventually 4 pups. They swarmed around the wolf like it was the female, but we learned later that it was the male.
>
> After about 30 more min., we decided to climb the hill where the female was in order to head farther N. along it to where we could see the entire den slope. We realized we would be spooking the female but took the risk. She moved and kept peering back at us from over the hill as we climbed. When close to the top, I looked back and there was the male wolf—the one we met at the WS and followed E. along the fiord. He was totally unafraid and 100′ away. We stood there and he walked around us but did not eye us warily; rather he just walked around nonchalantly like we were no big deal.
>
> We sat down and lay on our sides watching him, and he came up and sniffed Dean's boot. Then he came around behind me, put his nose toward the ground and came toward my rump.

What happened next will be covered in chapter 10, but we eventually continued to our new vantage point and glassed the four pups below. Brutus's little offspring seemed about 4–5 weeks old; their ears were still not prominent, but the pups were quite mobile. We later learned that the den was actually a fox den that the wolves had apparently just recently usurped as sort of a rendezvous site, which the pups could dive into if threatened but was too small for adults to enter.

So, finally in 2004, a few years after the unusually early winters, the Eureka wolf-prey system was recovering (Mech 2005). The wolves were breeding again, and that would continue, with Brutus "in charge" for at least the next few years. Dean and I never learned where the actual natal den was in 2004 but found it had not been in any of the dens I knew of. The 2004 fox-den rendezvous site was about 2 kilometers northwest of the original 1986 den.

During the next two years, the wolves did use the 1986 rock-cave den, producing three pups in 2005, with six adults present, and five pups in 2006, with seven adults present. The three years of sudden pup production, after seven years lacking it, yielded interesting data for trying to figure out how the Eureka wolf numbers relate to arctic hare and muskox populations (Mech 2007a). And, of course, those pup years also provided excellent opportunities for me to further my original interest in observing wolf behavior around the den.

The three years of new wolf denning differed considerably among themselves and not just because the number of adults and number of pups were different each year. In 2004, the denning situation was new, with quite-young pups (4–5 weeks old) living around that fox den. The next year, the pups were 7–8 weeks old when I found them and living in a rendezvous site rather than around any kind of den. (The reason I knew the pups had been born in the original 1986 den was because weather-station personnel had seen them there earlier.)

In 2006, the pups were about five weeks old and still using the original den. Thus, the interactions between the adults and pups during those three years differed among them. For example, the pups were nursing in 2004 and 2006, whereas they had already been weaned during the 2005 visit. At least some of the differences among these years were no doubt related to the 2005 study starting a week later than those of the other years, as in the High Arctic a week can make quite a difference.

The main commonality among all those years was Brutus being the putative breeding male. His identity each year was not in question. Each year he was seen, his behavior toward me and my colleagues was outstanding. Brutus's sniffing of my gloved hand in 2003 was followed one other year by his close, nerve-wracking inspection of my back (to be detailed later).

In 2005, there is this from my notes: "At some point about now, Brutus came to WM and tried to get under the front of his ATV for an apple core WM had dropped," WM being Walter Medwid, the former executive director of the IWC who had volunteered to accompany me that year. Walter is a consummate wolf and wilderness enthusiast who was totally in his glory there in 2005. He had accompanied me in 2001when neither of us saw even a single wolf, so this year was a bonanza for him. Not only were he and I daily observing both adults and pups, but at this moment Brutus had his whole head and shoulders under the front end of Walter's ATV. What more could a wolf buff ask for?

Then there was 2006. During 2006, in an exception to my practice of only a single colleague accompanying me, I hosted three IWC board members during this trip. As usual I had to teach each person the conventions I had learned for them to minimize disturbance while settled on their ATVs observing the wolves. Most important was not to dismount from the ATV, or at least to dismount on the side opposite the wolf.

Next in importance was to keep all gear on one's person or attached to the ATV, never on the ground. This year that latter convention proved its worth upon violation, much to Brutus's great delight. My field notes (July 9) tell it all: "6:50 p.m.—Brutus visited us with usual nonchalance, and we warded him off by saying 'No!' But he nonchalantly picked up Ted's pack lying on the ground behind his ATV and took it down to the meadow." There is still more to the story, but that will come later. The point here is that, while anywhere near the wolves, it was critical that all loose items be kept secured out of reach.

Besides that bizarre incident, 2006 was a productive year for the study, for I was able to watch the wolves during two more muskox hunts and gain insights into their hunting tactics that I had not been able to do before (see chapter 17).

Despite Brutus having frequented the Eureka area for four years, he eluded me in 2007. Weather-station folks had seen a pack of 10–12 wolves in mid-May, which may or may not have included Brutus, but the last time they saw Brutus had been May 20, with four other wolves. Dean Cluff, Valerie Gates, and I spent a week in July searching all the dens and areas I knew from the ground, and Dean and I even spent an hour in a helicopter to no avail. (Valerie is a philanthropist interested in wolf research who had funded various studies of mine as well as the International Wolf Center.)

But, lo and behold, some three weeks after our team left, weather-station personnel reported that Brutus showed up in the Eureka area with three other adults and four pups! Wouldn't you know? And where had they been? The pups were so young that it's hard to believe they came from very far away. On the other hand, "SuperPup" of 1990 had traveled 22 miles (35 kilometers) to a rendezvous site when about the same age.

:: 9 ::

South of Slidre

I finally solved the mystery in 2008. After much scouting, searching by helicopter, and ground tracking, I learned that, at least in 2008 and very probably in 2007, Brutus's pack had forsaken denning in the Eureka area. Instead, they had denned south of Slidre Fiord (fig. 1.3), some 12.5 miles (over 20 kilometers) southeast of the original den. Unfortunately, they had not trekked to the Eureka area in 2007 until after I had left or I would have seen their tracks in areas I had pretty religiously checked.

The new denning area was not only across Slidre Fiord, but it was also impossible to reach on foot or ATVs from my headquarters near the Eureka Weather Station. The station lay along the north shore of Slidre Fiord some 9 miles (15 kilometers) from the head of the fiord. Feeding into the head of the fiord was Slidre River, complete with wide mud flats on both sides. The wolves had been traveling from the den located south of the head of the fiord to the flats, swimming or fording the river, crossing the mud flats on the north side of the river, and then heading west along the north side of the fiord to get to the Eureka area, a round-trip trek of more than 25 miles (40 kilometers). They had been doing that frequently (Mech and Cluff 2009). In following them, I had backtracked from Eureka to the flats, where I then had to park and search the south side of the fiord via spotting scope.

This new finding posed quite a dilemma for me. If the wolves were to keep denning in this inaccessible area, that would prevent observing them any time I wanted to, which was the main advantage of this study. Still, I hated to give up on this interesting and productive investigation. Pondering that, it occurred to me that radio-collaring a few wolves would solve that problem. If I couldn't watch them around the den anymore, at least I could now start learning much more about their travels and other spatial information, especially if I used the new GPS radio collars.

GPS radio collars would send the data to my computer, so I could check them regularly from my office even during winter. Because I had been applying GPS radio collars to wolves in Denali and Minnesota as early as 1997 (Mech et al. 1998; Merrill et al. 1998), the idea was a natural fit. The GPS collars collected information about the animal's location, time, and date and transmitted the data to a satellite that relayed it to my computer. Why not try them on Ellesmere wolves?

As mentioned, I always hesitated to try radio-collaring Ellesmere wolves because I thought handling them might break the important rapport we had that allowed me to observe them up close. However, now after having spent so much time studying their behavior and being thwarted in trying to observe them around their den that was impossible to reach, I viewed my options as either giving up the project or starting to radio-collar the wolves.

For many years during my work with the Ellesmere wolves, I had imagined how radio-collaring one or two of them would help answer so many of my questions about them. Where did they go and what did they do in winter was the main question, but how far and wide did they travel during summers when I was unable to continue following them on my ATV? How often did they kill muskoxen, and where? Where were they denning in years when they were not in the general Eureka area?

Enter Brutus. Brutus did use the Eureka area again in 2009, but that year it proved to be his undoing as far as hiding his whereabouts and travels from me were concerned. It was clear that he and his pack were denning again in the inaccessible area south of the fiord. I observed at least 12 adult wolves in the pack from 4 miles (7 kilometers) across Slidre Fiord via spotting scope and binoculars (Mech 2010), and groups of up to 10 showed up in the Eureka area several times. Brutus would make an ideal candidate to bear the first radio collar of his clan.

Dean, who had accompanied me during 2004, 2007, and 2008, and I had planned to GPS-radio-collar at least two wolves in 2009. However, because of baggage loss and mishaps along the ad hoc flight chain, we ended up with only a single collar upon our arrival at Eureka. To deploy it on a wolf, we were equipped with both a dart pistol and a blowgun. The blowgun took a bit of practice, and Dean turned out to be the better marksman. All that remained was to get within blowgun range (1–3 meters) of a suitable wolf. Top priority for a suitable wolf was one of the breeders because they call the shots for the pack and would be most apt to remain in the area longest.

And guess which that long-lasting, top-priority wolf within range turned out to be? On July 8, 2009, at 3:45 p.m., five wolves were grubbing around at the Eureka landfill when Dean and I arrived, including Brutus and the breeding female. Brutus ambled right up to me and sniffed the day pack I had on my lap. When satisfied, the curious wolf lifted his head and took a few steps away.

Zzzzap! The dart from Dean's blowgun struck Brutus in the shoulder, in a good hit from a meter and a half away. The big wolf hardly noticed. Three minutes later, he stumbled a bit and in another minute he was down on his haunches. We attached the GPS collar, weighed and measured the beast, ear-tagged him, and checked his teeth for wear (Gipson et al. 2000). Dean and I agreed he appeared to be 8–10-years old. He weighed 90 pounds (41 kilograms), and he was the first wolf within 1,500 kilometers to be weighed and radio-tagged (Mech and Cluff 2011).

This milestone was also a turning point in that this long-running behavioral study was about to become an investigation into the movements and spacing of the Ellesmere wolves. I had had some notion about the approximate area my wolves had used during the island's short summer while the pups were growing and developing. However, neither I nor anyone else had any idea about how far and wide they traveled during the rest of the year and what they did then. If this GPS collar were to work well, it would be just a matter of time before we would find out what these animals do when it is –70°F and dark for months at a time.

While we processed and examined the tranquilized wolf in our own shadows, his packmates hung around and watched from varying distances, fairly nonchalantly and not really showing any concern but just lying around and coming by every now and then. It was not the kind of thing they were used to dealing with. Brutus stayed down until around 1830 hrs, when he was able to walk (or struggle) downhill, and he made his way to the very bottom of a hill. By 1930 he was able to walk uphill, appeared fully recovered, and walked normally.

At 2120 hrs, Brutus stood over a male packmate and pinned him down about 50 meters from us. For 6.5 minutes, the radioed wolf, with tail vertical, stood over, pinned, or held his packmate down by straddling or riding up on him (Mech and Cluff 2010). See http://www.youtube.com/watch?v=wIRVpLaCDS0.

Brutus paid no attention at all to his collar, nor did the other wolves, and eventually they all headed off together toward the den. However, along the way, each one split off on its own to hunt, with Brutus being the only one to cross the fiord and head back to the den. The breeding female had made off earlier and headed that way.

Of course, Dean and I were anxious to see how well the GPS collar was working, but we would have to wait for the location data, collected every 12 hours by an Argos Satellite, to be emailed to our laptop every 4 days. Meanwhile we spent as much time as we could trying to find and hopefully GPS-collar an adult female if possible, for the rest of our luggage had finally arrived with the second device.

During the next few days, we were unable to find any wolves. Meanwhile, Brutus's locations had come in via email, and it was clear the collar was working well. Seeing the unique data showing where Brutus had been for the past few days made us even more intent to get our second GPS collar on another wolf, any wolf.

Thus I hired a PCSP helicopter at Eureka to head to where the last location showed Brutus to be—well across the fiord and probably around the den. With a portable location antenna, we were able to home in on Brutus as the helicopter made its way toward the last known GPS location. Dean in the back seat was monitoring Brutus's signal, which led south to where we spotted a noncollared wolf lying on its side along a watery, green creek bed, and another around 200 meters upslope from it, standing. Then, a few hundred more meters upslope we found a den with three adult wolves near it and a pup with ears fully up (around six weeks old) that scrambled back down into the hole. The den looked like an old fox den with many small holes.

The chopper landed about 150 meters from the den, and a wolf sauntered over to inspect the strange bird. The pilot cut the engine. Dean got out to try to dart the wolf, but the animal would come no closer than around 10 meters—too far to dart even with a dart pistol. Dean was well experienced with darting wolves (and bears), and a dart here would have rendered the wolf safely unconscious just like the blowdart we had drugged Brutus with earlier.

We took off again and continued to track Brutus's signal, for we could see that none of the wolves around the den was wearing a collar. A few hundred more meters away, we finally found Brutus. There he was, lying with six others in a tundra crack around a meter deep and a meter wide, which provided the only shade for many kilometers around. It was a weird and unusual sight to see six wolves lying closely in a straight line along a trough in the middle of the tundra. Dean suggested that the chopper drop him off right away in the middle of the bunch so he could dart one of them.

That might have worked, providing a golden opportunity to get that second GPS collar out. It had to be a split-second decision, for the wolves would shortly flee from their tundra crack as the ship closed in. For some reason, however, this did not seem like a safe enough situation to me. Maybe it was the precipitous nature of the situation; maybe I wasn't sure how the wolves would respond to Dean. Maybe it was both. All I remembered about it later is that perhaps I should have let Dean try it, and if so, we might then have had a second set of GPS location data to view, ponder, and wonder about over the next year.

In any case, the wolves sprang away with their midday nap rudely interrupted. Thus, this new opportunity to collar another wolf was lost, and that happened to be the last opportunity for the duration of the trip. Brutus would remain the only collared wolf that year for thousands of kilometers around.

The GPS collar continued to work well, and for many months we were able to play armchair biologist. In our comfortable offices 1,250–2,500 miles (2,000–4,000 kilometers) away, we viewed Brutus's locations on our computer screens as he and his pack made its way around the Fosheim Peninsula and even neighboring Axel Heiberg Island. Occasionally Brutus and his bunch stopped by the weather station during the darkness of midwinter, and the hardy denizens there photographed them and emailed the shots to us. That way, we were able to determine that Brutus's pack comprised at least 20 members (Mech and Cluff 2011).

The collar recorded Brutus's locations at 6:00 a.m. and 6:00 p.m. each day. Even a pack of 20 wolves spends more than 12 hours around a kill feeding and sleeping. Thus, when we found that two or more consecutive locations were in the same place (a location cluster), there was a good chance that point represented a kill. Therefore, the next year, 2010, we would visit as many such locations as possible and search for the remains of whatever prey the wolves might have killed and eaten (Cluff and Mech 2023). Where several data points fell at the same location, those might lead to dens, one of which we hoped the wolves would be occupying that next summer.

That single GPS collar that Brutus carried around all winter turned up much interesting information about the pack's movements, information that I had been curious about ever since I first set foot on Ellesmere way back in April 1986. I had been suitably impressed then by my first exposure to the 24 hours of daylight that shone so brightly off the snowfields, and I also knew that many of the previous months had brought just the opposite, 24 hours of darkness. Whatever did the wolves do at that time of year?

No one had any idea. No one had even tried to learn. The closest winter wolf-movement study had taken place >1,400 miles (2,250 kilometers) south. For that matter, no one had studied the movements of any mammal at the 80th parallel north other than polar bears and Svalbard reindeer. Now, 25 years later, I finally knew and was able to broadcast that to the rest of the world (Mech and Cluff 2011).

Those findings just further whetted my appetite for more such information, so even as the data were beginning to display on our computers, Dean and I began making plans to collar more wolves and maybe even a muskox or two in 2010.

At the same time, personal considerations had been tugging at me. I had been getting older for some time, and my grandchildren were growing all too fast. I was already 7 years past qualifying for full retirement from my government agency but still healthy. However, I realized that if any severe medical problem were to arise on Ellesmere, it would be long before I could obtain adequate medical help.

The nearest medical specialist was thousands of kilometers away. It could take many days and $20,000 or so to even get a flight to Eureka to pick me up. When Canadian military forces are at Eureka for a few months, they do have a general-practitioner physician with them but hardly the equipment and supplies for any kind of major medical emergency. Personnel there are sometimes medevacked to Thule Airforce Base 330 miles (530 kilometers) away in Greenland. That probably would not be a choice for me.

Thus, at this major turning point in the project, I felt it best to start breaking in a successor for the research. Dan MacNulty, one of my University of Minnesota graduate students who had just completed his PhD degree, immediately came to mind. The Ellesmere study would make an ideal situation for someone just starting to set up their own research program, especially one based on GPS radio tracking. Dan jumped at the opportunity to join Dean and me in 2010 and received a National Geographic Society grant to do so.

An unexpected problem suddenly popped up, however. The Nunavut government had been issuing annual permits for the wolf study, and applications for those permits had to be approved by the Inuit Hunter Trapper Association (HTA) in Grise Fiord. Although approval was allowed for radio-collaring in 2009, the application for future collaring was denied.

The Inuit villages had suffered a long history of researchers who visited their territory, characterized by a lack of communication and invasiveness of approach. Thus, I had even hired an Inuk (one of the Inuit) to translate my 2009 report into Inuktitut to make sure the Inuit could read it. Still, there had been a history of abuse and trauma at the hands of representatives of about every colonial authority figure—governments, RCMP, churches. Thus, it isn't surprising that Western science and its practitioners have sometimes been greeted with suspicion. The ripple effect of all this was that, while our 2009 collaring had been permitted, further collaring was not approved. Sometimes issues arise on another project with other researchers or under different circumstances and are applied as a precaution to similar projects. Regardless, the permit was not approved for 2010.

Dean, Dan, and I hoped that a way could be found to overcome this problem, but we also realized that our 2010 trip might mark the end of this long-running research. Nevertheless, we still had that summer to help us add to the project's data. Each of us had been looking forward for many months to checking on the ground the information our computers had been displaying via the GPS tracking of Brutus and his pack.

And that we did. Unfortunately, old Brutus had died in April 2010, apparently of cancer, so when my colleagues and I arrived in July, there was no longer a collared wolf we could follow. Still, we had the information about all the twice-per-day Brutus locations since collaring our prize wolf in 2009. We knew that when wolves kill large prey, it takes them a while to devour it and then sleep nearby. I had documented that in detail for my pack in 1989 (Mech 2011). That meant that when two GPS locations 12 hours apart were the same—the location cluster mentioned earlier—that could well indicate a kill. It was a technique I had been using in Minnesota a few years before (Palacios and Mech 2010).

Thus in July, we three eager biologists began our hunt for kill remains using a portable GPS device to home in on location clusters. Our lack of a permit to radio-collar any more animals did not forbid us from checking the locations where last year's data showed wolves possibly had made kills. Excitement increased each time as we gradually closed in on the target location while scanning the area ahead for the first sign of a bone, patch of hair, or other piece of prey remains. We three Ds searched 9 of 53 wolf-location clusters from the ground and 26 more by helicopter; three of these clusters were searched both from a helicopter and from the ground (Cluff and Mech 2023).

Those searches led us to 17 sets of carcass remains (1 caribou and 16 muskoxen) and to two active wolf dens (≥ four adults and ≥ four pups at one and ≥ two adults and five pups at the other). One den was several kilometers southeast of the end of Slidre Fiord. The other was a few kilometers northeast of the head of the fiord.

Brutus had visited both these dens during the estrous period in late March/early April 2010 but at no other time since October 17, 2009, for one den and since November 17, 2009, for the other. Autumn den visiting had only been reported once before (Thiel, Hall, and Schultz 1997), but I knew of no reports of wolves visiting dens during the estrous period. On Ellesmere, the dens would usually be snow-covered during March and April, so the significance of a visit then is hard to explain and might just be a coincidence. However, visiting both dens 11 miles (18 kilometers) apart at an interval of just two days and during the estrous period seems more than coincidental when that was the first time either was visited in several months.

The 2010 study turning up the two new dens and all those kill remains by checking Brutus's GPS data really excited us. Those discoveries showed the potential of the GPS collaring technique to add new information to the Ellesmere study even after my long series of summer investigations. We were keen to apply the technique there more fully in future years.

I, however, would not be part of that operation. Because of the considerations mentioned above, I had decided to say my goodbyes to Ellesmere Island, knowing that for sure 2010 was to host my last stint there. The island had been my summer home for 24 years, and the wolves, my primary companions each summer. We had now demonstrated the value of GPS-radio-tracking wolves there and had launched Dan as the new potential leader of the research.

It would not be an easy transition for Dan and Dean, however. My funding for the annual trip itself, which had always constituted most of the project's cost, had come from my ongoing US government research project. How would Dan manage such costs? His 2010 trip was supported by a special National Geographic Society grant, but where would all that required funding come from for an annual trip?

In addition, GPS collars are expensive, and contrary to our blow-darting of Brutus from an ATV for a pilot project, any future study of several wolves would have to rely on darting from a helicopter. That additional expense for collaring would be hard to scrape up. Furthermore, even if funding could be found, how would the men obtain approval from the Inuit to do any radio-collaring?

Thus, there was a lull of a few years in the study.

What eventually boosted the project was, of all things, a TV documentary. After the three documentaries that had already been made, word was out among the world's wildlife documentarians that the place to film wolves was Ellesmere. However, the filmmakers also knew that to overcome the logistical issues there and to actually find the wolves there to film would be major obstacles.

Still, over the next couple of years, a few such filmmakers contacted me wanting to make more documentaries. Each time, I gave them a brief overview of the problems and the possible solutions. The clear way out for them to find wolves to film was to cooperate with a scientific team that was studying radio-collared wolves. That in turn would require the team to obtain approvals from the Inuit and the filmmakers to arrange to work with the biologists needing to do the collaring. Funding for the biologists, everyone's travel, the collars, and the helicopter time would all have to be covered. A win-win for all.

I begged off the collaring jobs myself but sicced the film folks on Dan MacNulty and Dean Cluff, who were only too willing to get involved. Later Morgan Anderson, a biologist interested in studying caribou and wolves on Ellesmere, joined the team.

Morgan, a Government of Nunavut employee, had met often with the Inuit about Peary caribou management. The communities frowned on collaring caribou or muskoxen to track mortality and survival but recognized the importance of this information in land-management decisions that they were critical in evaluating and informing. However collaring wolves was more palatable, and Morgan and the communities saw the utility of following the predators to learn the pressures on the prey.

By the time Dan and Dean reached out to her, Morgan had already been flying surveys, planning research, and sharing results with the Inuit. Morgan is not Inuk, but she did live in Nunavut full time, in Igloolik, and as such was a local contact with some experience and knowledge of the multi-faceted challenges faced by Nunavummiut (people living in Nunavut). She opined on the best lake ice for tea (although with grudging deference to Grise Fiord's glacier water), swapped stories of the best loot found at the dump, and was happy to take the oft-repeated blame for every unseasonable blizzard as a weather witch (a dubious honor that haunts her to this day). With a clear set of objectives informed by community priorities and discussion about methods, risks, and results, our study became something Grise Fiord was actively involved in shaping and delivering. The permits went through.

Finally, June 2014 marked the new phase of the Ellesmere studies, which continued to July 2018. I was ecstatic. Now I could resume studying the Ellesmere wolves, albeit from afar. Sure, it was nothing like visiting the island, searching out the den, and befriending the wolves each summer. No more Brutus sniffing my glove. No pups trying to untie my bootlaces.

Nevertheless, given my long curiosity about what those same wolves did over winter, where they went, and how they might have interacted with other wolves, merely being able to sit at my computer and check each day on their movements, not just for a few weeks in summer but year-round, was an excellent consolation. And often when I saw the points on the map that showed this information, I could envision those places, for I had been there. In a sense, it actually brought me back to where I had spent so much of my time following the wolves in person.

Among five filmmakers over the four years, funding allowed my colleagues to attach 10 GPS radio collars to four adult male and six adult female wolves representing possibly as many as six packs. By that time, the collars had improved greatly, allowing biologists to remotely reprogram the location-acquisition rates and to drop the collar off the wolf when the radio's transmission life was over.

Yes, at least four packs and possibly six used the general Fosheim Peninsula region, within 30 miles (50 kilometers) of Eureka and the original den I had found so long ago. Much information besides just the actual locations of those packs was also collected, data on kills they made, some counts of their prey, their genetics, and how many pups they produced each year.

Of course, all of these high-tech methods of studying the Ellesmere wolves, with helicopters, dart guns, blood-samples, and GPS radio collars, was a very far cry from my original approach using merely a notebook and pen. In a very real way, the evolution of study techniques during this research had recapitulated the development of these techniques over the period of many decades in the rest of the world in a very short time.

Still, the motivation was the same: collect as much information about wolves as the situation allowed. In the following chapters, I will discuss that information, from the nitty-gritty details about newborn wolf pups emerging from the den to the overview of a pack of 20 or more full-grown wolves making their way across thousands of kilometers of ice and snow during months of constant darkness, bitter cold, and shearing wind.

:: **10** ::

Cast of Characters

As has been apparent along the way, certain of the Ellesmere wolves I studied featured prominently in my observations. To keep them straight and for my notes, I named them, usually according to their various characteristics (Mech 1988a, 1995b), contrary to my studies elsewhere, in which I numbered them. The individuals that remained from summer to summer in the Ellesmere pack I basically lived with each year were generally the breeders, formerly referred to as alpha wolves in earlier publications (Mech 1970, 1999).

Other biologists have suggested that we need to know much more about individual wolf personalities, for like probably every other organism, each is different, and those differences can critically affect each individual's behavior. I noticed this many decades ago with the wolves I studied while they attacked moose on Isle Royale when I stated (Mech 1966, 62), "one wolf seemed to be more aggressive," though I had no way of knowing it was always the same wolf. Earlier, Murie (1944, 28) had described a prominent member of his East Fork wolves as "lord and master" of the pack.

In relation to wolf attacks on humans, Linnell, Kovtun, and Rouart (2021, 3) also felt "an urgent need to learn more about the behavior of 'bold' or 'fearless' wolves," and others have documented how individual wolves have affected ecosystem services (Bump et al. 2022). The International Wolf Center even published a book about individual wolves in which various wolf researchers recounted tales of their favorite study animals titled *Wild Wolves We Have Known* (Thiel, Thiel, and Strozewski 2013). Thus, it is no surprise that as I worked so closely with the Ellesmere wolves, I would see each wolf as a different personality.

Those that stood out were Mom, Whitey, Explorer, Left Shoulder, and Brutus (table 10.1). Mom was the mother of the pups from 1986 through 1989. She was a thin, obsequious individual in 1986, often with her tail between her legs and, although as tame to me and my companions as were most of her cohorts, she was still jumpy and shy. Generally, she was ragged looking, with shedding fur clinging off many parts, and she seemed to rank lower than one of the other females in 1986 and 1987. In 1988, the only other adult female present was Whitey, almost certainly a yearling daughter of Mom and that deferred regularly to her. Mom seemed more self-assured then. In 1989 she raised pups again and appeared "businesslike," but by early August she was being dominated by her daughter Whitey.

Mom remained with the pack in 1990 but as the grandmother to the single pup, as daughter Whitey became the breeder. A year later, Mom remained but apparently had been kicked out and had to wheedle her way back in as Whitey's two pups grew and developed. Mom was nowhere to be seen during summer 1992.

Whitey, apparently born in 1987, began dominating Mom, the wolf I assumed to be her mother, in August 1989 and took over the breeding role in 1990. First breeding at the age of 3 years accords well with what I learned about the wolves I was studying in Minnesota (Mech, Barber-Meyer, and Erb 2016). Whitey seemed to be a more secure individual, although that could have been due to her not having any older female to deal with other than her mother which she had displaced as breeder.

Table 10.1. Prominent adult wolves observed near Eureka, Ellesmere Island, during this study, 1986–2010

Name	Sex	Tenure (summers)	Remarks
Mom	Female	1986–1991	Breeder 1986–1989
Whitey	Female	1988–1996	Yearling 1988; breeder 1990–1996
Explorer	Female	1993–2000	Yearling 1993; paired nonbreeder 1998, 2000
Left Shoulder	Male	1986–1996	> 1 yr. old, 1986; breeder 1988–1996 (no pups 1993, 1995)
Brutus	Male	2003–2009	Paired 2003; breeder 2004–2009; unknown 2007

As mentioned earlier, Whitey's mate appeared to be Left Shoulder, a member of the 1986 pack that would have been at least a year old in 1986 and thus could have been an older sibling. Ellesmere wolves, living on an island and at the edge of the species' distribution, are known to be genetically less diverse than those more centrally located in the species' distribution (Carmichael et al 2008, Frévol et al. 2023). Thus, such inbreeding would not be so surprising.

"Explorer" was assumed to have been born in 1992 and was paired in 1998 with a male that was unusually afraid of humans. That animal almost certainly was not a local, for I had never encountered a local wolf that was so intimidated by a human. Rather the wolf must have originated at least 90 miles (150 kilometers) away, the closest location where wolves are sometimes hunted. However, wolves disperse so far that Explorer's mate could have come from much farther away.

Explorer herself, however, was quite tame and self-assured. She was the wolf at which in 1993 I had to clang the pots and pans to dissuade her from grabbing the sleeping bag that I had been airing out on the tundra near my tent (chapter 7). She was also quite a hunter, for I saw her and her mate kill a cow muskox in 1998 in about 5 minutes, and Explorer sometimes grasped the creature by the nose (chapter 15). On the other hand, Explorer never did have pups during the 2 years when I studied her as a member of a pair, and I did not encounter her again after the second year (2000).

Of the males, two stood out, with Left Shoulder's tenure as the breeding male being the longer, lasting from 1988 to 1996. Originally part of the first pack I found in 1986, he stood out because of a fist-sized wound on his left shoulder, presumably from being gored by a muskox. During 1986 and 1987 he seemed to be second in command and big brother to the pups, chewing bones with them and patronizing them in various ways. Then, in 1988, he became the breeding male.

Living so long and having such a long tenure as a breeder is outstanding in itself for any wolf anywhere, although I did know an individual like this during my Minnesota studies. Female wolf 5176 lived 11 years and remained with her natal Perch Lake pack through her last year, presumably as its breeding female (Mech and Hertel 1983; Mech unpublished). Left Shoulder lived so long that he began losing his voice and his howl sounded like a lion's roar (chapter 20). Although he was still able to catch arctic-hare leverets when at least 13 years old, he had to rest for long periods afterward (Mech 1997b).

Of all the wolves I got to know on Ellesmere, "Brutus" dominates. He even became the subject of one of my chapters in *Wild Wolves We Have Known*

(Mech 2013, 216). "When the wolf I came to know as Brutus deftly sniffed my gloved hand, I knew right away that he was not your regular, 'run-of-the-mill' wolf of the High Arctic," I wrote then.

Brutus was far more than tame; he was blatantly bold. As noted earlier, in 2005 Brutus was the wolf that shoved his head and shoulders under the front of an ATV that my associate, Walter Medwid, was sitting on. The wolf was seeking an apple core that Walter had dropped, and Brutus was intent on snagging it. That being Walter's first close encounter with Brutus, this former executive director of the International Wolf Center was suitably astounded. I, on the other hand, recalled Brutus's sniffing of my glove two years before so I didn't flinch at Brutus's awkward attempt to get a little dessert.

I did flinch a bit from Brutus a year before, however. Here, from *Wild Wolves We Have Known*, is what I wrote:

> One time he even frightened me, the first time in my then 46 years studying wolves I had been afraid of one. In 2004, when my associate Dean Cluff (Department of the Environmental and Natural Resources, Northwest Territories, Canada), and I were searching for Brutus' den, we took a break, and Brutus found us. As Dean and I were lying on the tundra, I was perched on an elbow, and Brutus sniffed my boot. This was the first time I had ever been off my ATV around Brutus, and I wondered how he would react to the 2-person, 400 pounds (182 kilograms) or so of human biomass right down on his level. Sniffing my boot was fine, but when Brutus ambled around behind me where I could not see him, I had second thoughts. I recalled someone's German shepherd several decades before that had nipped at the back of my clothing all the way up to my hair and had actually nipped some of that. I also recalled in 1987 being within a few yards of a muskox calf as several wolves tore at it until it collapsed (also shown in the *White Wolf* video). Now with Brutus right behind me, for a minute or 2 I became truly frightened.
>
> "You know, Dean, for the first time in my life I am truly afraid of a wolf," I uttered to my companion. I had thought that maybe Brutus would tug at the day pack on my back, and I might jump or whirl around, possibly triggering some predatory response. I knew only too well what a single bite from Brutus could do.
>
> Dean offered little consolation as he lay 10 feet (2.5 meters) from me watching Brutus behind me. "All I have is my leatherman tool" said

> Dean. [We had left all our gear on our ATVs hundreds of meters away.] That was more than I had, although of course no small tool of any sort would have been sufficient to ward off an attack. Each second Brutus stayed behind me seemed endless until he finally moseyed back around to my side where I could see him. He then strolled some 20 feet (5 meters) from us, lay down and howled. Dean recalls glancing at me for eye contact to acknowledge "how awesome is this—to be so near a wolf on the High Arctic tundra while he is howling!?" My fear was all for naught, and I ended up feeling foolish. (Mech 2013, 216)

Brutus is also the wolf that made off with the day pack of another of my summer companions in 2006, mentioned earlier and described later in detail (chapter 12) and the one which we darted and fitted with a GPS radio collar (chapter 9). After he died, weather-station personnel retrieved his carcass, which showed he was a victim of cancer. His body was then sent to a taxidermist to be mounted for educational display. Brutus had been an especially bold wolf, but he taught us much, and he now suitably guards the Eureka Weather Station as a bold mounted representative of the area's well-studied wolf population.

:: 11 ::

Ellesmere Dens

The entire Ellesmere project had depended on my finding a wolf den during my first summer visit, in 1986—not only finding a den but finding one in the general region of Eureka's landing strip that I could access reasonably easily. And one where the adults would tolerate my presence close enough to observe them and their pups, at least via binoculars or a spotting scope. Given what I know now about the proportion of summers when wolves there den in that region (about half the time), chances in 1986 were about fifty-fifty. As of this writing, that is quite a sobering thought.

As one would expect for a region as far north as this, the area is underlain with permafrost, so wolves cannot just dig a hole in the ground like in so many other places farther south. Rather, they need to resort to rock caves, crevices, or special landforms such as well-drained hillocks or cutbanks that are free of ice or frost and dry out early in spring. The team that followed me found several inactive dens filled with ice, which is probably partly why the rock cave I found was so consistently attractive and maybe why in some years the wolves need to just use pits in which to bear their pups.

In many areas, wolves tend to den toward the center of their territories (Taton 2023), but in some places proximity to a ready food source seems to override that tendency (Ciucci and Mech 1992). On Ellesmere, and no doubt in other regions where topography, weather, and other physical features restrict availability of suitable sites, any kind of shelter must be sought.

As it turned out, not only was the den I found suitable for my studies, but it also was so suitable for the wolves that they used it in 8 of the 11 summers during which I observed pups between 1986 and 2006. During the other summers either the wolves I had been studying failed to produce pups or they denned so far away that they did not frequent the usual Eureka study area. As indicated earlier, I found wolf-chewed bones at the main den that were as old as 162 to 983 years. The cold aridness of this area makes it possible to find and analyze bones this old, unlike further south.

Even in 1990, when the pack's single pup was born in a dusty pit dug into the top of a cut bank about 4 kilometers from the traditional rock-cave den (Mech 1993b), the wolves eventually moved the pup to that den. The first move came some 14 hours after my companion and I found the pit and began observing, when the wolves moved the pup 1.2 kilometers to the shelter of an underslung boulder about the size of a vehicle. I felt there was a good chance that the wolves, although as tolerant of me as usual, had moved the pup because of my presence. A day later, I watched as the breeding female carried the pup away from the rock and to the traditional den some 1.6 miles (2.6 kilometers) away, which the wolves used for the rest of their denning period that year.

During two years, the wolves did use another den some 3 miles (4.8 kilometers) from the rock cave but just for part of the time after the pups were born elsewhere. This den was a hole dug under a large rock outcrop on a ridge high above Slidre Fiord. It had probably started as a fox den that the wolves enlarged, and it opened into a miniamphitheater-like area surrounded by rock outcrops. The first time was in 1991, when Whitey produced two pups in the same pit den as in 1990. She moved them several times to other pits she dug as she went, but she was never satisfied until she secreted them in this den with a view of the fiord (below). She kept them there at least until August 8, my last day there that summer.

Whitey's 1991 behavior protecting her pups was of especial interest because it demonstrated how concerned she was about them. I stumbled on them on June 14 after finding the main den unused and then searching elsewhere for where the den might be. I headed to the pit where Whitey had borne her single pup the year before, just wondering what the pit would look like after a year of weathering. "To my great surprise," I wrote, "there were 2 live wolf pups! They were mottled gray and about 2 wk. old (1730 hrs). Big head, tiny ears. Eyes barely open or not open." Whitey had again produced her pups in the pit den!

My companion and I got right out of there, realizing that being that close to the young, highly exposed pups could upset the adults. We headed across the valley where we could view it from a distance. That was a good thing, too, for 10 minutes later, Whitey showed up. She sniffed the area closely and then lay in the pit and nursed the pups.

The pups in the pit were exposed to the weather whenever Whitey was gone or not lying in the hollow with them, although they survived rain and snow for many days (Mech 1993b). Nevertheless, Whitey seemed insecure about where they were. During the next few weeks, she moved them twice to new areas and several times when within a given area, digging pit after pit (table 11.1). She also seemed wary of allowing her mate near them, and even at times seemed to be wary of my partner and me even though most of the time she was as tolerant as any wolf in previous years.

Table 11.1. Key observations of the 1991 wolf pups born in a pit den, Eureka, Ellesmere Island, Nunavut, Canada

Date	Time (hours)	Observation
June 15	2150	Whitey moved pups 50 meters south and lower down bank to new pit and stayed in there with them.
June 17	1034	Whitey dug three nearby pits and put the pups in a new pit a few meters below where they were. All her short wanderings away from the area could be in search of better spots.
June 18	0249	Pups in pit 2.
	1207	Whitey left pit 2, but one pup crawled up edge of pit, and Whitey carried it to bottom of bank and across valley, then right back up to pit 2 and nursed it until 1212 and stayed in pit.
	1420	She picked up pup, dropped it, and picked it up three times, took it down slope, across valley, and then back to pit.
	1721	Whitey carried pup down slope, disappeared west.
	1737	Whitey ran back along her back trail toward the pit.
	1739	Picked up pup 2.
	1745	Heading west with pup and disappeared over ridge.
	1819	Whitey sniffed pit 2, scratched at it; sniffed at pit 1 a few seconds, and left.

(continued)

Table 11.1. *(continued)*

Date	Time (hours)	Observation
June 19	1616	Pups in new pit on hillside about 0.8 kilometers from original pits. Several boulders nearby and Whitey checked rock crevices about 150 meters south of them and returned. As Blackmask (BM) returned toward pups, Whitey ran to the pups and lay with them for the third time when BM headed toward them.
June 20		Snow day; unable to travel.
June 21		There are now two new pits, one below a rectangular rock and one above it.
	2146	Whitey moved pups 6 meters to a rock ledge. A pup lying on Whitey suddenly rolled off and rolled 2 meters downslope. It then returned to Whitey, who seemed unperturbed.
	2340	Whitey left pups near rock ledge, checked other rocks and dug among them; still seems insecure about where she has the pups.
June 22	1539	BM to pups in lower pit and lay with them, but Whitey rushed over and BM left and lay 9 meters south. Whitey then lay 2 meters from pups.
	1741	BM to pups; they whined. Whitey explored rock piles 150 meters away, ran over, submitted to BM, picked pup up, and put it 0.6 meters back into pit, moved 1.5 meters away and nursed pups from 1744:40 to 1748:30. Then spent much time checking rocks and digging.
	1838	BM went to the pups, and Whitey displaced him. She nursed pups 1844:30 to 1849:18.
	1852	Whitey carried pup 30 meters toward new pit. When BM went to second pup, Whitey carried the first back to him, then to new pit 150 meters away, and BM followed. Whitey carried second pup over and lay with pups several minutes, then sniffed old pit and surroundings for 1.5 minutes, back to new pit and 300 meters beyond and disappeared, 1913.
June 23	1740	Whitey with pups in the newest pit.
June 24	1901	Whitey headed to pups behind a large boulder 40 meters northeast of her last pit. Mom with them till then. BM not there.

Date	Time (hours)	Observation
	1910	Whitey left pups and lay 6 meters away.
	2029	Whitey appeared; pups come 6 meters from behind rock and nurse from 2032:10 until 2038:10. Whitey led them behind rock. One strayed from behind rock; she carried it back, then checked other rocks.
June 25	1953–1958:39	Whitey took the pups 7.5 meters above rocks, nursed. Pups climbed on her and followed her. She carried one back down to rocks, and the other followed.
	2008	Whitey went back above the rock and slept 7.5 meters away.
	2100	Whitey carried the pups, singly, to rocks and slept with them.
June 26	0400	Whitey and pups still near the rocks.
June 27	0150	Whitey digging around new rocks but finally slept with pups.
	0934	Whitey moved the pups about 15 meters uphill and nursed them.
June 28	0209	Pups wandered while Whitey and BM slept; one got 61 meters away.
	0700	Whitey seemed disturbed by one of us walking.
	0712	She barked and howled unlike ever before for 3 minutes. Then she headed to pups, led them 0.8 kilometers and lay with them. As they fell behind and howled, she returned for them.
	0836	Whitey and pups still 0.8 kilometers away in small gully and peeking at us with just eyes showing, a sure sign of her being disturbed. We left at 0845.
June 29–July 2	1533	Not much pup viewing but noticed their ears were standing—still in same pit and rock area.
July 3	0055	Whitey arrived, went straight to pups and led them away 0.8 kilometers, possibly disturbed about us.
	0115–0426	Pups in new spot, but Whitey and the pups highly tolerant of us within a few meters.
July 4	0045	Pups in same new place as yesterday. Whitey explored new rock piles, returned to the pups and nursed them at 0331.

(continued)

Table 11.1. *(continued)*

Date	Time (hours)	Observation
	0347	Whitey carried pup 150 meters to rock pile.
	0351	She took the second one there; did not return to check if there were any more left.
July 5–8		Pups same place as yesterday.
July 8		All wolves gone and to new den 1.5 kilometers southwest of last pits. Enlarged fox den at base of rock ledge.

Note: Whitey's behavior indicated concern that the pups were unsafe in such pits and continued until several days later when she secured them in a hole in the ground, likely an enlarged fox den.

In 1992, the pack was using the original 1986 den when I arrived on July 2, but 2 weeks later, they moved to the fiord den, where they remained until at least August 5. I had no idea why the wolves moved. In fact, moving the pups to the fiord den seemed foolish, for only 4 days before, the wolves had killed a bull muskox in the opposite direction from the rock-cave den. By moving the pups, the adults had made the distance to haul food from the carcass to the pups 3 miles (4.8 kilometers) longer!

The only other summers between 1986 and 2006 when the Eureka wolves bore their pups in dens other than the above were in 1989 and 2004. In 1989, as detailed earlier, the den was 14 miles (22 kilometers) east of the traditional den and was not found until July 23, so that might not even have been the natal den. I only saw it for a few minutes because Rolf Peterson had found it while I was back in Minnesota, and when I returned, the wolves had already moved. According to Rolf's notes, this den was on a low ridge among some rocks, and it had a "lawn-like" spread below it where the pups played and slept a lot. I took a quick look at the empty den and then concentrated on finding where the wolves had gone.

In 2004, I never did find the natal den. Rather Dean Cluff and I found the pups on July 10 when 4–5 weeks old at an old fox den. Because none of the holes was large enough for an adult wolf, the pups must have been born elsewhere. Both the traditional den and the fiord den housed fox pups that year, so neither could have served as the wolves' natal den. Nevertheless, the natal den was probably in the general vicinity of the traditional den, because where they were found was within 2.5 miles (4 kilometers) of that den.

After 2006, it was not until 2010 that the Eureka wolves denned again on the north side of Slidre Fiord that was easily reached by ATVs. For reasons known only to the wolves, but possibly to take advantage of a new herd of arctic hares (chapter 16), the pack had begun denning to the south of Slidre Fiord, as described above. Recall that the Eureka Weather Station and airstrip lay on the north side. At the head of the fiord the Slidre River entered, complete with its extensive mud flats ready and waiting to trap any ATV daring to try to cross there.

Wolves, however, crossed with impunity, punching into the mud blatant tracks and trails that teased my companions and I who sought their dens. It was not until the end of my 2008 stay that I figured this out, and with spotting scope on the south side of the river then managed to pick out tiny white wolves dotting the distant tundra on the other side. Because a nursing female was one of the eight adults I observed visiting the Eureka area that year (Mech and Cluff 2009), I knew the pack had pups, but there was no way I could observe them. That is when Dean and I decided that the next year, 2009, we would start using GPS collars.

Having so collared Brutus early during our 2009 stay, we were able to home in on him south of the fiord via a helicopter flight and locate his pack's den, an old fox den with several holes midway up a gentle slope. Although the pack was still using the Eureka area, their route around the end of the fiord to the den was 19 miles (31 kilometers) or, in a straight line across the fiord, 10 miles (16 kilometers). Earlier during the season, they might have taken the shortcut across the ice of the fiord, but while Dean and I were there, the pack was taking the land route around the end of the fiord.

In 2010, I had the benefit of much information from Brutus's GPS collar about where Brutus traveled during the rest of 2009 and early 2010, and in checking out the clusters of location points indicating where he spent extra time, we found a new active den. On the evening of July 11, the first GPS cluster the helicopter flew to turned out not to be the remains of a kill but rather a den with at least four adults and at least four pups. Brutus had visited that site on October 17, 2009, and on March 31, 2010. The den was on a gentle slope and had several holes.

Another GPS location cluster showed that Brutus had visited a second den on September 17, 2009. This den was back on the north side of the fiord again, about 14 miles (22 kilometers) east of the traditional den, not far from the 1989 rendezvous site. Like the other den Brutus visited, the area is much dryer with sandier soil, and the den was a hole dug into the ground, complete with a den mound much like dens farther south. In 2010, at least two females and 5 pups denned there. During many hours of watching this den from July 8 to 16, I never saw a male there, similar to the situation in 1989.

Nevertheless, Brutus had at least visited both dens in 2009, thus implicating him as perhaps having something to do with the wolves occupying them in summer 2010. He didn't die until April 2010 so could even have fathered the pups in either or both of the 2010 dens. In any case, by September 27, 2010, weather-station folks counted some 14 pups and five to seven adults visiting their area (Eureka Weather Station Weekly Status Report, September 27th, 2010).

For 2011, I was unable to get much information about wolf denning in the Eureka area except that a September 27, 2011, email from the Eureka Weather Station reported that their personnel had counted some 25 wolves, including seven pups passing by the station. They were also seen there on October 22, 2011.

Also of interest is that, although I had no information about where the wolves denned in 2011, in 2012, student aide Alex Vnukovsky from Fleming College (Peterborough, Ontario) at the weather station found that at least three adults and five pups were again using the original 1986–1988 rock cave den that they had temporarily abandoned after 2006.

The only den I got a chance to examine in any detail was that rock cave. The tunnel was wide enough for an adult to easily fit in for at least 1.5 meters and continued for at least a meter longer. It did not slant down much but rather ran straight back under the rocky ridge. Both Murie (1944) and Clark (1971) described the insides of wolf dens in detail.

Generally, at lower latitudes after the pups are about 8 weeks of age, wolves leave their dens and settle into less secure areas on top of the ground. Most intraspecific mortality of pups takes place when the pups are only weeks old (Douglas Smith et al. 2015), so the safety of a hole in the ground is no longer as necessary as the pups grow. These new homesites above ground, or "rendezvous sites," are sometimes near old dens or are deep in thick vegetation that still offers some protection. The pups remain there, often in a big ball when asleep, awaiting the return of the adults to feed them just as when they lived in dens.

Rendezvous sites can be close to natal dens or many kilometers away. Wherever they are, their main function is to be a place for the pups to reside while they grow and develop, and the adults expend their energy to procure food for them. As soon as the pups are large enough to accompany the adults, usually around October or November, they abandon rendezvous sites and become nomadic with the adults within their territory.

Because Ellesmere wolves are usually born in early June, and I usually had to leave the area by early August, I learned little about their rendezvous sites. During a few summers, the wolves did abandon dens while I was still there, with the most dramatic being the 22 mile (35 kilometer) move that 1990s pack of three made with its single pup mentioned earlier and detailed later. Of the five summers when wolves did move their pups to rendezvous sites, twice the sites were around old fox dens, where pups might be able to squeeze into some of the multiple holes found there, and once at a secondary den. The other times, the pups were left out on the open tundra (table 11.2).

Once the pups were moved to rendezvous sites, they were mobile enough that they often tried to follow the adults as they left for their daily hunts. Sometimes the adults also moved the pups to other sites temporarily, but during the periods I was observing, usually no later than early August, the adults always moved their growing offspring back to the rendezvous sites.

Table 11.2. Basic wolf rendezvous site (RS) information, Eureka, Ellesmere Island, Nunavut, Canada

Year	Earliest known/possible date at RS	Distance from natal den (km)	Characteristics of RS	Remarks
1987	July 30	0.4	Open tundra	Might have been there sooner
1989	July 27	10.0	Fox den	Last known date at den 7/26; not at den 7/30; found at RS 8/4–8/10
1991	August 1	35.0	Secondary wolf den	After birth in pit den and moves to pits elsewhere
1992	July 16	5.0	Same secondary wolf den as in 1991	Until at least 8/6
2004	July 13	–	Fox den	Probably were there sooner
2005	July 9	5.3	Open tundra	Probably were there sooner

Note: We did not find the natal den in 2004.

:: **12** ::

It's All about Pups

Unlike Clark (1971), I made no special effort to record the growth, development, and behavior of the pups, nor did I make observations in any structured fashion that might have featured the highlights of pup development. Rather I recorded all my observations about all pack members at all times of day ad lib on notepads. I then elaborated on them within 24 hours in journal form.

To link observations to pup age, it would have been best to have known when the pups were born each year, but that was always before I arrived. The birth date can vary from year to year at any given latitude because a wolf's estrous (heat) period is about two weeks long and the start of an individual wolf's estrous period can vary by a month (Kreeger 2003). Thus, in any one location, over the years pups could be born over a duration of at least a month.

However, I was able to approximate the birth date of pups in 1991 based on the age of eye opening. That age in captive wolf pups raised by various workers is the 11th to the 15th day (Mech 1970). In addition, the eyes of pups born on a known date on Baffin Island (about 69° N) opened at 13 days of age (Clark 1971). The earliest I found pups on Ellesmere was June 14, 1991. That is when I found the two males born in the pit den. Because their eyes were barely opening, I estimated they were about 2 weeks old, indicating they were born about June 1. The latest any could have been born there was about June 10, based on data from 12 summers (Mech 2022).

Therefore, to estimate the age of pups of the unknown-age litters, I could consider the 1991 litter as known-aged. I could then use other key events for these pups at various ages, such as ears barely standing, to age the other litters. For example, if the ears of the 1991 pups were just barely standing when 4 weeks old, then I assumed that when the ears were just barely standing in an unknown-age litter, those pups were probably also about 4 weeks old. Then I could use other observations in that second litter to match with those of other litters to age them.

This method assumes that the pups of each litter were fed reasonably consistently each year. Because their main source of sustenance for their first several weeks is nursing (Packard, Mech, and Ream 1992), that was probably the case. However, the actual ages of the unknown-aged litters could vary by a few days from those I estimated. Thus, I was able to relate various stages of development to various pup ages (table 12.1).

Table 12.1. Observations of wolf pups made around dens near Eureka, Ellesmere Island, Nunavut, Canada

Estimated pup age (days)	Observation	Date	Year
13	Tiny ears; eyes barely open or not; my impression was they could not see. Much pushing around with their heads against mother's body; legs not well developed. They are mostly head and body.	June 14	1991
17	Eyes open; climbing to top of pit den. Female moves both pups.	June 18	1991
20	Can walk. Ears still small.	June 21	1991
20	Pup first emerges; can barely stand; ears visible.	July 2	1996
20	Emerged from (the rock cave) den (after female carried it 2.8 kilometers from a pit den first seen on June 23).	July 2	1990
21	Big head; little ears.	July 3	1996
22	Larger nose; ears; urinates by itself.	June 23	1991
23	Sprawly; can barely stand.	July 5	1996
24	Pups ate regurgitant for several minutes.	June 25	1991
25	Ears still down.	July 7	1990
25	Ears not prominent.	July 10	2004
26	Howled.	June 27	1991
27	Eats grass.	July 9	1996
28	Walks but waddles; ears standing; chewed on leveret and could pull fur from it; 3 meters from den.	July 10	1996

(continued)

Table 12.1. *(continued)*

Estimated pup age (days)	Observation	Date	Year
29	Pup ate regurgitant for about 4 minutes.	July 11	1990
29	8 meters from den.	July 11	1996
29	Ears about to stand.	July 1	1994
29	Ears first start to stand.	July 3	1992
30	Pups sleep in front of den; ate regurgitant.	July 12	1996
30	Ears half standing.	July 12	1990
31	Ears up; tiny; playing.	June 24	1987
32	Mother licks pup anogenitally.	July 14	1990
32	Ears up; short; legs short; nose blunt.	July 4	2006
33	Urinating alone; ears standing.	July 5	1994
33	Ears up but not protruding above top of head; pups look kitten-like.	June 21	1988
34	Playing.	June 22	1988
34	Pups 50 meters from den.	June 23	1988
35	Ears prominent; drank water from stream.	July 17	1996
35	Yearling licked pup's bottom.	July 7	1994
35	Carrying meat with fur; tail up.	June 24	1988
36	Can't find food when within 15 centimeters; ear tips not totally up.	July 18	1996
37	Cached food twice near den.	July 19	1990
38	Pup caches hare part.	July 10	1994
39	Eating hare meat.	July 10	1991
39	Appear kitten-like.	July 21	1996
40	75 meters from den.	July 22	1996
41	Pup caches bloody fur.	June 30	1988
42	Urinating on their own.	July 13	1991
44	Cached 225 meters from den.	July 26	1996
46	Nose pointed; ears up.	July 10	2005
48	Pup travels 12.5 kilometers to muskox carcass during a round trip of 39 kilometers lasting 12 hours, 19 minutes, including a 26.5-kilometer trek in 5 hours.	July 30	1990
52	Pups initiate group howl; urinate alone.	July 11	1988

Source: Mech 2022.

The number of pups I observed each year varied from 1 to 6 and averaged 3.3 (table 12.2a) if years when no pups were produced are not included in the mean. During years that lacked pup production, one could argue that the Eureka pack merely denned somewhere far from where I checked, and that could be true. However, if the Eureka pack is defined as the wolves that used

Table 12.2a. Number of adult and pup wolves observed around dens near Eureka, Ellesmere Island, Nunavut, Canada

Year	Adults	Pups
1986	7	6
1987	7	5
1988	4	4
1989	8	4
1990	3	1
1991	3	2
1992	2	3
1993	5	0
1994	4	1
1995	3	0
1996	2	2
1997	3	0
1998	2	0
1999	no data	no data
2000	2	0
2001	1	0
2002	0	0
2003	3	0
2004	2	4
2005	6	3
2006	7	5
2007	5	0
2008	8	at least 1
2009	10	at least 3
2010	2	5
2010	4	4

the Eureka area during the summer, then the finding that they did not reproduce during several years is probably accurate. This is because when I concluded that no pups were produced it was because I either determined that the wolves using the Eureka area included no females with visible nipples or that no wolves used the area at all for any particular summer. Even wolves that denned as far away as 10 miles (16 kilometers) straight-line distance (19 miles [31 kilometers] travel distance around the head of the fiord) still showed up around Eureka during summer; for example, those in 2009 and 2010 (Mech and Cluff 2009).

Both the litter sizes I found (table 12.2a) and the average litter size (3.8) that Morgan Anderson et al. (2019) turned up in the broader study area later (table 12.2b) were similar to those of the entire Canadian Arctic Archipelago, 3.9 (Miller and Reintjes 1995). However, they were higher than those in Greenland (2.0; Marquard-Petersen 2008) but below most areas farther south, which vary from 4.0 to 6.5 (Mech 1970; Mech and Boitani 2003). Wolf-pup litter sizes are very much a function of the amount of prey, with more pups per litter where prey biomass is higher (T. Fuller, Mech, and Fittz-Cochrane 2003).

Table 12.2b. Number of adult wolves and pups found on the Fosheim Peninsula, Eureka, Ellesmere Island, Nunavut, Canada, in 2014–2018

Pack	2014	2015	2016	2017	2018
Eureka	15–16	13 (3)	10 (11[b])	4 (3)	3+
Rock Den[a]				2 (5)	
Gibbs Fiord (Axel)	7 (9)	9	6 (6)	6	
Vesle Fiord	3 (3)	5			4 (2)
Hot Weather Creek	2 (0)		2 (5)	2 (3)	
Cañon Fiord	6 (3)		4		6 (4)
Mt. Lockwood	5 (3)		2 (1)		
Bay Fiord	2 (3)				
Wolf Valley	4 (4)				

Source: Anderson et al. 2025.
Notes: Numbers in parentheses indicate pups.
[a] Pack apparently broken up and pups at three dens. (This year is not included in calculation of the mean.)
[b] Uncertain affiliation with Eureka pack because the den wasn't active for most of the years of the study.

Basically, the arctic wolf pups grew and developed about the same as wolf pups elsewhere. That is not surprising because the timing of wolf mating is related to latitude (Mech 2002), as are the seasons and the reproduction of prey. Thus, wherever and whenever wolf pups are born, it is the season when prey have produced their young too, so there is usually plenty of food available for them.

However, how quickly captive wolf pups develop during their first few weeks can vary. For example, the ears of a pup raised by the International Wolf Center in 2021 were not yet standing at 4 weeks of age, whereas the ears of those raised by Mech (1970) were already standing by that age.

This variation probably also explains differences in reported ages of pup emergence from dens. Stanley Young and Howard Goldman (1944), with the US Fish and Wildlife Service, stated (without documentation) that pups emerge at about 3 weeks of age, and Mech (1970) merely cited them. Captive pups in Indiana emerged from their den when 3 weeks old (Klinghammer and Goodmann 1987).

Complicating this issue further, Kim Clark (1971) found Baffin Island pups appearing at the den entrance at 9–11 days. He is the only researcher who observed newborn free-ranging wolves emerging from their den after he thought he knew when they were born. The pups emerged even before their eyes were open, assuming Clark was correct in judging when they were born. The only information about his basis for determining the pups' birth date follows: "7th June is the assumed birthdate of the 1969 litter based on observations of the bitch on May 24, June 7, and June 9th" (Clark 1971, 78). This statement raises the question as to specifically what information Clark based his assumption on about when the pups were born.

Without further information, one can wonder how valid Clark's assumption was, especially because some of his observations at various ages do not seem to accord with what is known about very young pups in general. For example, one pup in Clark's study that he assumed to be 13 days old ventured 1.8 meters from the den; this does not comport well with observations from hand-reared captives (Mech 1970). Clark also felt that weaning began at 19 days and was complete at 30 days, 5 days younger than Schonberner (1965) reported for captives and weeks earlier than we found for Ellesmere pups (Packard, Mech, and Ream 1992).

On the other hand, Clark's (1971) pups' eyes opened at about 13 days of age based on when he thought the pups were born, which accords well with the date of eye opening that is pretty well documented (summarized by Mech

1970 and discussed earlier). I was fortunate in 1991 to have found two pups at just about the age of eye opening born in the pit den (described earlier and by Mech [1993b]), and I could then describe them at that age of about 2 weeks: "Tiny ears; eyes barely open or not; my impression was they could not see. Much pushing around with their heads against mother's body; legs not well developed. They are mostly head and body" (table 12.1).

This description does not accord well with that of Clark (1971), who saw the 13-day-old pup crawl 1.8 meters. Still, 4 days after I found my 1991 pups (described earlier), they were able to climb to the top of the pit den such that the female had to move them to prevent their leaving the pit. Four days of growth brought a great deal of change. The best assumption then about Clark's (1971) assessment of when his pups were born and his estimates of pup ages is that he was reasonably accurate and thus correct about the pups emerging from the den at 9–11 days. Nevertheless, this age of den emergence is far different from what Murie (1944) found.

Thus, given all this uncertainty and conflicting information, it may well be that, contrary to our assumptions, the development of free-ranging, young wolf pups is far more variable during their first few weeks, or at least that their age of first appearance outside the den is highly variable.

On Baffin Island, wolf-pup teeth began emerging at about 17 days (Clark 1971), and those of captive wolves are mostly erupted by 3 weeks of age (Mech 1970). The Baffin pups began eating solid food when 17 days old (Clark 1971), although the pups' main source of nourishment comes from nursing. On Ellesmere, I recorded pups aged 24, 29, and 30 days first eating regurgitant (Mech 2022). From this age on, adults routinely regurgitate chunks of prey to their offspring, which then eagerly gobble them up. As I will discuss later, this means of feeding seems to generate important social bonding as well.

In 1996, a pup I judged to be 28 days old chewed on a leveret carcass and could pull fur from it, but during week 5, the 1998 pups still mouthed small meat chunks a long time before swallowing. At an estimated 72 days of age, a 1988 pup was able to chew and swallow the hind foot and leg bones of an arctic hare.

However, fueled daily by their mother's milk and whatever regurgitant they can manage to chew, the pups grow and develop quickly. As they do, their ability to compete for regurgitant and snarf it down accelerates, thus propelling their growth. Long ago, Erkki Pulliainen (1965) identified three growth periods. During the first period, from 0 to 14 weeks, the pups grew maximally, at an average rate of 42.3 ounces (1.2 kilograms)/week for females and 52.9 ounces (1.5 kilograms)/week for males. The second period, 14–27 weeks, saw rapid growth averaging 28 ounces (800 grams)/week for both sexes. During the last period, 27–51 weeks, growth was slow, with females growing 1.1 ounce (30 grams)/week and males, 7.1 ounce (200 grams)/week.

Long before reaching their first birthday, the pups are procuring food on their own, pouncing on insects and no doubt eventually lemmings. As discussed earlier, I did get a chance to watch as a 28–33-day-old pup seemed to learn how to deal with a live, injured, arctic hare leveret that a yearling carried to it (Mech 2014).

Finally, in autumn the pups join the adults, becoming nomadic with the rest of the pack and joining them in killing hares and muskoxen. I never got to observe that period because I was always gone by then. However, elsewhere even 4–7-month-old pups orphaned from their parents survived (Fritts, Paul, and Mech 1985), and 6-month-old pups can kill prey on their own (Bass 1992). Nevertheless, the pups are still growing and developing, and they will continue to be fed by their parents even as yearlings if there are no new pups (Mech 1995a).

Even when there are new pups, the adults sometimes do deliver food to yearlings, as I saw on July 19, 1994, when the breeding male brought in a leveret estimated at about 2.2 pounds (1 kilogram). Whitey did not rush him that time, but Explorer did and snatched the hare from him and promptly devoured it. The pack produced only a single pup that year, and there were only two yearlings in the pack. Thus, the amount of food needed for either Whitey or the pup was much less than usual, so the yearlings also benefited from the parents' foraging.

Despite their long sleeps and food deliveries to pups, wolves do find time to otherwise interact with the pups around the den and with each other. Although so many of those interactions revolve around food, such as caching prey pieces, digging up caches, and stealing food remains from each other, there sometimes still is time for play.

The youngest pup I saw playing was at an estimated 31 days of age (table 12.1) on June 24, 1987. My notes simply say, "The pups played around the front and W. side of the den." However, Clark (1971) watched pups he estimated at 18 days of age wrestle and chew on each other. About pups estimated to be about 24 days old, Clark (1971, 79) wrote, "Pups run, jump, wrestle on hindlegs, push with forepaws. One wins, attacks a third." As noted earlier, Clark's pups might have developed much faster than the Ellesmere pups or there may be an issue with his or my estimated ages, for some Ellesmere pups were "sprawly" and hardly able to stand at an estimated 23 days of age (Mech 2022).

In any case, wolf pups at least a month or so old play regularly. From a technical standpoint, one might argue about just what kinds of behavior should be regarded as play, especially in pups. However, anyone can identify most kinds of play merely from their own experience with pets and children: chasing, wrestling, tumbling, jumping, stalking. pouncing, tossing items, feigning fighting, playing tug-of-war, and similar behaviors.

Even adult wolves indulge in the same kind of activities, much like dogs. Play helps make muscles strong and stretched, keeps various kinds of behaviors practiced and at the ready, juices social interactions, and probably even increases an individual's self-confidence. Many of the playful motions and behaviors are really the same as or similar to those that every wolf will use to search for, approach, and attack prey.

Whether wolves or other animals actually play for enjoyment in the same way as humans is an open question. Certainly, numerous observations of various species of mammals and birds give the impression that they enjoy playing, as anyone who owns a dog can attest. All kinds of scientists have studied play and speculated, experimented, and hypothesized about its evolutionary function or value, although with little agreement.

The bottom line may actually be that many creatures do actually enjoy playing, as Vanderschuren, Achterberg, and Trezza (2016, 86) documented in rats. They found that several opioid neurohormone systems play a prominent role in modulating social play, especially dopamine. Their technical summary that "in sum, social play behaviour is the result of coordinated activity in a network of corticolimbic structures, and its monoamine, opioid and endocannabinoid innervation" evinces that play is no simple matter. Recent work involving tickling experimental rats identified the periaqueductal gray area in the midbrain as a play center (Gloveli et al. 2023). Bekoff (2007, 2024) has maintained for years that animals actually enjoy play.

Whatever the reason for, function of, or value of play, the Ellesmere wolves engaged in it over and over, sometimes for just minutes but at other times for hours. I documented it regularly, at times even at unusual moments. For example, on July 14, 1988, Grayback, a yearling male, had been gone for 28+ hours, yet when he returned, "he still plays with the pups for the next forty minutes." Probably the longest bout of pup play that I recorded, for 3 hours and 45 minutes, also took place in 1988. That included the pups traveling playfully for a few hundred meters between a nearby rendezvous site and the den.

The adults themselves sometimes burst into a frenzy of behavior that seems to have no function other than play. In 1994, when there were four adults but just a single pup, such a frenzy developed around the den. On July 19 the whole pack was awake but just lying around. At 2124 hrs the breeding male headed to the pup and lay with it. A few minutes later, a subordinate male joined them, and the pup began playing with the two males for about 15 minutes, then joined the two females and slept for about 20 minutes. The young female, Explorer, strolled over and played with the pup vigorously for 10 minutes, then ran to the breeding pair excitedly. All three approached the other male, and suddenly the whole group, including the pup, charged exuberantly down the valley below the den toward our tents, some 400 meters away. They all sniffed around the campsite area, howled, and headed back to the den.

What was this all about? My partner and I had camped in that spot since June 30 and the wolves had visited our site and traveled past it numerous times, both while we were there and while we were gone. The wolves must have known just about everything there was to know about the place. Still, suddenly, in a fit of play, they decided to check it out again during a short outing. Why? A little squirt of dopamine perhaps?

So the pups played among themselves, the adults did the same, and the adults and pups mixed it up in every imaginable way. This activity could happen spontaneously with any wolf starting it or it could precipitate when one of the pack approached, especially one of the parents. In 2004, the breeding male, which had been around the den area for at least two hours, headed to the pups on the den mound at 1712 hrs; all four flocked to him and swarmed around him, over him, etc. He licked the underside of a few of them and let them play with him until 1735. Then he headed to the female and both slept.

The next year, I watched as three females started a free-for-all with the pups, stimulated by the return of a subordinate female which regurgitated to the breeding female and three pups. When all the feeding was over, all three females and all three pups wrestled, chased, and played, mostly wrestling, for 20 minutes.

Sometimes a novel object really gets the gang going. Over all the summers that I spent with the Ellesmere wolves, I saw such novel toys brought in as a dead fox, a seal skin, and a plastic bag of apples from the weather-station landfill. Still, the best example of a novel toy was the day pack that Brutus stole from one of my companions in 2006. Described earlier, it deserves another mention because that day pack with its peanuts, M&M candies, and other human goodies was no doubt the finest toy the Ellesmere pack had ever found. "All the wolves [seven adults and five pups] converged on it and ripped part of it open (we later learned) and they ate the candy, peanuts, cookies, and scattered the hat and plastic bags, peed on it and rolled in it for several min., carried it around, etc. STU'd [made a standing urination] on it."

It doesn't take a very novel item to precipitate play, however. Even a simple bone or a piece of hide can so serve. These July 16 notes from my first year on the island presaged many a time during the remainder of my studies on Ellesmere: "There was much playing with a hare leg bone today by the pups and the adults and digging up caches and chunks of meat and carrying them around and disgorging food to the pups. The adults stayed around all day and much of the playing took place 12:00 p.m.–3:00 a.m."

Even a single wolf can find a way to play all by itself at times. In 1993, I watched a yearling female spend 5 to 10 minutes digging up a lemming. She then flipped it a couple of times, let it run, and then for 5 more minutes played "wolf and mouse" with it before finally chomping it down.

: : **13** : :

Many Mouths to Feed

Neither Murie (1944) nor Clark (1971) described much about wolf-pup nursing. Therefore I and my 1988 colleagues, Jane Packard and Bob Ream, did a special study of it. We observed nursing with the Eureka pack from June 21 through August 1, 1988, when the four pups were some 5–10 weeks old (Packard, Mech, and Ream 1992).

The Ellesmere pups all nursed simultaneously, both while the female was standing or lying, but not in the same locations each time. The particular configuration of the 1988 den was the cave under the end of the base of a tall rocky ridge. That allowed the wolves, including the pups, to frequent either side of the ridge, putting them out of sight from the other side. Thus, my two associates and I had to find different viewpoints to keep the female in view. We also had to spell each other for various periods, with two of us watching from different viewpoints while the third slept. That way, our group was able to observe around the clock.

Both the mother and the pups initiated suckling at various times. The mother began it by approaching and whining softly near the pups, which often were sleeping. The pups initiated suckling by approaching their mother, often when she returned from somewhere but occasionally when she was sleeping. I always found it fun to watch the pups as they swarmed around their dutiful mother, each pushing against the other to gain a free nipple.

Mom ended nursing by walking away or turning her rear away from the pups. If a pup broke contact with the nipples prior to the mother moving away, we considered the suckling as terminated by the pups. This interpretation was justified because sometimes the mother seemed to sense that the pups were finished. For example, when one or two pups left her belly and moved near her head, she often looked back toward the pups, took a few tentative steps, and then walked away if pups did not persist. In such cases, she did not appear to stop suckling against the pups' interests.

The pups suckled for about 3 to 5 minutes at a time when they were 5 weeks old but decreased to just a minute or so when they were 8 weeks old. They fasted between nursing bouts for periods a bit less than 5 hours when about 5 weeks old. As they began weaning when 9–10 weeks old, they fasted for up to 15 hours (fig.13.1) This behavior was contrary to the earlier weaning that Schonberner (1965) and (Clark 1971) reported.

As with nursing initiation, either the pups or their mother also ended the nursing. However, the proportions of the actions by mother and pups starting and ending the nursing shifted as the pups aged. The pups' persistence with nursing decreased with age, as did the proportion of nursing that their mother initiated. During the pups' sixth week of age, Mom initiated most of the nursing. That declined gradually to none in week 10 when single pups initiated suckling attempts. (During weeks 6–9 the whole litter had typically initiated nursing.)

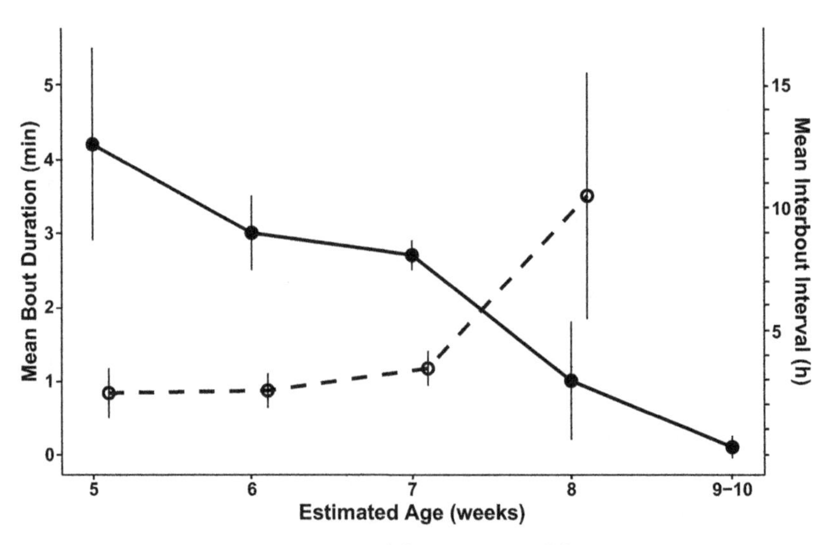

Figure 13.1. Changes in mean duration (●) and intervals (o) between suckling bouts relative to estimated age of pups in the Ellesmere wolf pack. Vertical lines show 95% confidence intervals (Packard, Mech, and Ream 1992).

During weeks 6 and 7, as the mother began to wean her pups, three or four continued to attempt to suckle after she terminated it. However, they persisted less in weeks 8–10, and as their mom ended the nursing, she sometimes muzzled the pups and often regurgitated meat to them.

The pups did not necessarily agree when their mother would not allow them to nurse, as I observed with a different mother on July 2, 1994: "0025 The pup tried to nurse, but Whitey snapped at him (came to E. side into my view) and pup bit at Whitey's tail and leg and Whitey snapped back, sat down, and tried to thwart pup's aggression. There was an element of anger on the part of both the pup and Whitey, not play. The pup at times jumped up and down on all fours in a way I haven't seen before. Both then slept together."

By their sixth week, the pups had started stalking and pouncing and the "standing over" posturing common in wolves (Mech 2004). They had been snarfing up regurgitant from the adults for weeks, and when adults opened hare carcasses, the pups readily fed from them. The usual response when an adult arrived at the den with stomach full or mouth carrying food was for the pups to rush it, thrust their muzzles to the adult's mouth and lick it, a behavior known as "licking up." The adult would then regurgitate, and the pups would "wolf down" the regurgitant.

We did a study of this routine during six summer periods from 1988 to 1996 and found the following based on 168 regurgitations during 114 adult wolf arrivals:

> Usually when a foraging wolf returned to the den area it was eagerly met by the pups and often by all the members present. When it regurgitated, sometimes more than one class [ex. pups and yearlings] of pack members fed.
>
> In 76% of the bouts, arriving wolves regurgitated where they were met, and in another 11% they were followed by the recipient(s) for 10–50 m before regurgitating. At other times they were followed up to 800 m. Wolves regurgitated only once in 61% of the bouts, twice in 24%, and three to five times in 12% over periods of 5–35 min. Often subsequent regurgitations followed solicitation by recipients, but at other times the regurgitator spontaneously re-approached recipients as though inviting solicitation. (Mech, Wolf, and Packard 1999, 1193)

At times the pups had to beg repeatedly before getting an adult to regurgitate. For example, field notes on July 26, 1996, read: "1038–1041 [hrs], the pups licked up to the male 4–5 times, and he kept moving and lying down but they kept after him, and at 1041 he regurgitated to them."

The pups quickly learn to be very competitive and aggressive with their food. In mid-July 1986, I saw one grab a lump of meat about the size of its head and run off about 150 meters to eat it. Another one growled back at its father which was growling while guarding a chunk of meat; the pup then snatched a piece from its parent and ate it several meters away.

It was easy to conclude that the pups' routine of following the regurgitators and then immediately getting fed quickly trained them to follow the adults, which they would then do for the rest of their time with the pack. They would also continue to depend on those adults for food, sometimes even as yearlings until eventually they would learn to hunt well enough themselves to assume the initiative during hunts and provide for themselves.

The pups grew quickly as they scarfed up the spoils the adults brought home, whether a few juicy pieces or a whole pile. I was intent on trying to figure out how much the adults were regurgitating each time; of course that no doubt varied. Further, because the pups instantly usurped the slimy mass of meat chunks that the adults afforded them, there was no way I could measure it. Still, the amounts were enough that one day (July 11, 1990) I recorded that the breeding male had regurgitated enough that Whitey and her pup fed on the heap for a full four minutes. After all, a wolf's stomach can hold at least 22 pounds of meat (10 kilograms).

One indirect way I could get at this problem, however, was to weigh the amounts adults regurgitated into caches. Based on three such amounts I dug up, the average was 44 ounces (1.25 kilograms). Each bout of regurgitations included 1–5 upheaves. I assumed that the 1.25 kilograms each also applied to the regurgitations that the pups consumed directly. That was an untested assumption, but would mean the adults delivered 1.25 to 6.25 kilograms of meat each time they arrived. Because the average proportion of regurgitations per bout was 1.5, that would average 1.9 kilograms per arrival. (These figures differ slightly from those reported erroneously in Mech, Wolf, and Packard 1999, which were 1.25–7.25 kilograms delivered per arrival and averaging 2.2 kilograms.) Generally, when a pack member came back from a trip away from the den, their return heralded some kind of food delivery. The main objective—and probably the only one—for being away from the den and pups was to procure food. My two companions and I observed at the den for 168 continuous hours from July 12 to 18, 1988, when the pups were an estimated 7 weeks old and still nursing. We saw a total of 34 adult wolf returns to the den, or about one wolf return per 5 hours on average. The wolves delivered food during 80%

Plate 1. The main den around which Mech made most of his observations was a cave under the base of a sandstone ridge.

Plate 2. The rocky tundra in front of the den was an ideal place for the pups to explore and play.

Plate 3. In 1990 only a single pup was present when Mech arrived in early July.

Plate 4. Peary caribou, the smallest subspecies, inhabited Ellesmere but in very few numbers. Photo by Morgan Anderson.

Plate 5. Arctic hares formed a substantial part of the wolves' diet, especially in summer.

Plate 6. When a pup strayed too far from the den, it was not unusual for Mom to merely pick it up and carry it back.

Plate 7. During nursing time, the pups scrambled to ensure they were well positioned.

Plate 8. Big brothers and sisters were favorites for pups to cuddle with.

Plate 9. Tugging on an adult tail tip was sometimes a handy way to gain attention.

Plate 10. By accompanying wolves on their travels, Mech could estimate their speed and observe their daily activities. Photo by Dean Cluff.

Plate 11. When a muskox herd flees, the wolves immediately chase them.

Plate 12. Dean Cluff accompanied Mech during several summers and helped observe the wolves.

Plate 13. A hearty meal awaits wolves when they finally kill a muskox.

Plate 14. The adult wolves spent much time lolling around with the pups. Photo by Nancy Gibson.

Plate 15. Wolf pups love to snuggle and play.

Plate 16. Adult wolves are often off chasing arctic hares or hunting muskoxen.

Plate 17. To beg food from adults, pups "lick up" to them, triggering them to regurgitate food.

Plate 18. When four to five weeks old, pups can hear, see, and smell and begin to explore their world.

Plate 19. The average litter size of Ellesmere wolves is four pups, similar to the wolves of Greenland.

Plate 20. Ellesmere wolves must make good use of every muskox they kill.

of their arrivals. (In a later study during 1,759 hours of watching the den over 5 summers including 1988, the mean time between individual wolf returns was about 8 hours; see Mech and Merrill 1998).

The trips during the 168-hour period of continuous observations probably did not include any from local kills of muskoxen or the wolves would have made several more each day transporting pieces of the burly beast back to the pups.

With 21–25% of the deliveries, the wolves only carried the food in their mouths, usually consisting of arctic hares or their parts. Once, in 1992, they brought a fish of about 2.2 pounds (1+ kilogram). In 43–54% of the deliveries, they carried nothing in their mouths but did regurgitate, and in another 12–18% they both carried food and regurgitated it. When the wolves delivered food by regurgitations, they often regurgitated multiple times (Mech, Wolf, and Packard 1999), and during the 168 hours of continuous observations, they regurgitated a total of 29 to 34 times (table 13.1).

The breeding female ("Nipples" was Mom's tentative name in 1988) returned to the den and pups the most times and delivered food the most times by a considerable margin both during this period (table 13.1) and most of the time (Mech, Wolf, and Packard 1999). My field notes from July 6, 1988, exemplify this:

> At 1938 [yearling female] Whitey returned with a chunk of meat that I'd guess was muskox. She went to the pups at the den . . . and dropped the meat there, then took the pups off and regurgitated to them at 1939 and led them W. out of my sight for a few minutes.

Table 13.1. Number of times Ellesmere Island wolf-pack members returned to the den/pups and provisioned the pups during 168 hours of constant observations during July 12–18, 1988, Eureka, Ellesmere Island, Nunavut, Canada

Results	Breeding female	Breeding male	Female helper	Male helper
No. returns to den	14	6	6	8
Empty	2	0	0	4+1?
Only carrying food	4	1	0	1
Only regurgitated	4 + 1?	3	4 + 2?	1
Carrying food/regurgitated	2 + 1?	1 + 1?	0	1
Total no. regurgitations	11 + 2?	8 + 1?	6 + 2?	4

Note: Question marks indicate occasions when the action was out of sight and so only inferred from the behavior.

> At 20:09, Nipples returned from the E. . . . She had nothing in her mouth but the left side of her neck was covered with fresh blood. She went straight to the pups and took them away from the den and Whitey and led them 150 m E of the den through the hillside and regurgitated to them when the pups tried to nurse. They did not eat all and followed Nipples about 30′ W and tried to nurse again. She led them back to the regurgitant twice about 20:18.
>
> At 20:34, Nipples regurgitated again and the pups ate it.

Although food deliveries went mostly to the pups (Mech, Wolf, and Packard 1999), all pack members except the breeding male received food from other pack members around the den (table 13.2). This lack of feeding of the breeding male accords with the findings from 1,102 feedings of captive wolves

Table 13.2. Numbers of regurgitations by wolves of various classes, Eureka, Ellesmere Island, Nunavut, Canada

	Breeders[a]				Auxiliaries				
Year	Female		Male		Female		Male		Total
1988	46[b]	(47)	18[c]	(18)	21	(21)	13[c]	(13)	98
1990	2	(10)	15	(75)	3	(15)	–		20
1991	3	(38)	4	(50)	1	(13)	–		8
1992	6	(50)	6	(50)	–		–		12
1994	0		8	(40)	3	(15)	6	(30)	17
1996	4	(31)	9	(69)	–		–		13
Total	64		60		28		19		168
Mean[d]	10	(27)	10	(27)	7	(19)	10	(26)	9
Mean[e]	6	(24)	7	(31)	5	(20)	6	(24)	6

Source: Mech, Wolf, and Packard 1999.

Note: A few regurgitations in a given bout may have been missed because of occasional obstruction during observations. Values in parentheses are percentages.

[a] No significant difference between male and female over 6 years (Wilcoxon's signed-rank test, $z = -0.67, p = 0.5$).

[b] Value contributed to significance of χ^2 goodness-of-fit test ($\chi^2 = 26.49, p < 0.001$); higher than expected by chance (Freeman-Tukey deviate, $z = 3.69$).

[c] Values lower than expected by chance ($z = 12.961$).

[d] Average number of regurgitations per wolf per year.

[e] Average number of regurgitation bouts per wolf per year.

over a seven-year period (Fentress and Ryon 1982). The breeding males, it seems, just must fend for themselves along with trying to feed everyone else (chapter 18). One exception I noticed, and it seems to have been the only one, came on June 22, 1991. The breeding male stole a very dried-up hare from Mom and even prevented Whitey from taking it away from him.

When we examined the regurgitations of these Ellesmere wolves over several summers, we learned that individual wolves regurgitated up to five times per bout. The overall ratio of regurgitations per bout was 1.5. Pups were more likely to receive regurgitations (81%) than the breeding female (14%) or auxiliaries (6%). The breeding male regurgitated mostly to the breeding female and pups, and the breeding female regurgitated primarily to pups. The relative effort of the breeding female was correlated with litter size (Mech, Wolf, and Packard 1999).

As for the specifics of what types of food the adult wolves were delivering and regurgitating to the pups, I had little information other than what I could observe directly, basically arctic hares and muskoxen. Probably I would have missed such items as birds and lemmings. It is interesting that on Baffin Island, according to scat analyses, the pups were eating significantly more smaller prey (14%) than were the adults (5%; Clark 1971).

:: 14 ::

Food Caching

The wolf's feast-or-famine lifestyle sometimes means that some wolves will go hungry for long periods. Although a well-fed adult wolf can withstand starvation for weeks or months (Mech 1970), pups cannot. In most wolf range, wolves bear their pups during the season when their main prey produce their own offspring, which are the most vulnerable prey of the year. Ellesmere is no exception. Thus, during most summers, wolves manage to procure enough leverets and muskox calves to feed their pups fairly regularly. Of course, this ability varies by the year. Nevertheless, during every summer when I found pups on Ellesmere (14 summers with a total of 49 pups), they all survived for at least the time I was there, which was usually in late July or early August.

An important behavior that helps buffer the highs and lows of an irregular food supply is food caching. Wolves are master cachers. Just like dogs bury bones, wolves bury just about anything extra that they value, which may be anything from a hare ear (July 31, 1996) or a muskox penis (July 12, 1996) to a full-blown muskox leg (July 9, 1998). A wolf in Denali National Park cached two caribou calf legs and the creature's head in separate caches (Murie 1944). Mostly, they bury meat. And every wolf does it, starting with pups as young as 37 days old (table 12.1).

I dug up three arctic-hare leverets that the wolves cached in 1996, and excerpts from my notes follow: "leveret #1 weighed 1.2 kg and was 35 cm. It was buried head first. It had about 10% grey fur in mid back; lev #2 weighed 1.32 kg, was all white and 41 cm nose to tail and also buried head down; the [third] cached hare weighed 1.45 kg and was all white."

Further, wolves cache both the food they carry in their mouths but also that they carry in their stomachs (Mech et al. 1999). The masses of six regurgitations I recorded from five caches, including those discussed earlier and reported by Mech, Wolf, and Packard (1999), follow: 88 ounces (2.5 kilograms) (two regurgitations that included 30 chunks averaging 80 grams) and 38.8 ounces (1.1 kilograms) by the breeding male and 49 ounces (1.4 kilograms) by the breeding female; and 23 ounces (0.65 kilograms) and 0.66 kilograms—all walnut-sized muscle chunks—by a different breeding female.

Caching can involve merely tucking a chunk of something into a handy crevice, or more formally, and most often, it involves digging a hole in the ground, dropping the item into the hole, or regurgitating a load into it, and then, with upward nose pushes, covering the cache with the loose soil the cacher had dug out. Although a wolf might decide to cache something at any instant, probably the most significant caches are made more deliberately, for example, after a kill, when individual wolves might travel for kilometers to the place where they decide to cache something (Mech and Adams 1999).

There could be several reasons for such long-range deliberate caching, but one reason can be because wolves tend to be secretive when caching as if they want to prevent their associates from knowing where their cache is. In Denali Park, Murie (1944, 112) noted one instance where a wolf, while heading to cache something, took to backtracking, walking in a stream, and circling around, seemingly trying to throw off track any creature trying to follow it to its cache. Even in well-fed captive packs, wolves tend to avoid caching when other wolves are present (Townsend 1996).

Ellesmere wolves were also secretive cachers. I noticed this during my first observation of the wolves feeding on a kill, on July 15, 1986 (Mech 1988a). My field notes indicated how impressed I was with "the degree of secretiveness about caching. Each animal goes off by itself and watches to see that no other individual is nearby. They seem almost paranoid about the caching." In 1988, a pup about 10 weeks old started to cache part of a hare but stopped when two other pups started following it. This behavior reminds me of an observation McIntyre (2020, 88) made in Yellowstone. A pup watched another carry away a chunk of food, zigzag around out of sight and cache it. The watching pup then scent-tracked the caching pup through a maze of other tracks right to the cache and stole the food.

Wolves do not always stray so far away from a kill to cache. While the pack of seven fed on three calves they had just killed in 1986, some only traveled

150–200 meters away to cache. Pieces cached then included lungs as well as hunks of meat and hide.

Sometimes around the den, caching can be a kind of more open free-for-all. From my notes again: "7/26/96 1033- A pup cached a chunk of something some 225 m NE of the den. A few minutes later Whitey [the breeding female that year] dug up the cache and carried the piece to the pups. A pup then tried caching it again elsewhere, and Whitey came over and grabbed it away. A few minutes later, a pup took the chunk about 75-m away and cached it (1054). At 1109, Whitey went to the cache and dug up the piece and ate it."

As to how wolves choose a specific spot in which to cache something, I could not figure that out. On July 8, 1998, for instance, when Explorer was caching after having gorged on a fresh muskox kill, she was always intent and searched around, zigzagging and looking around a great deal for several minutes to choose her spots. The only other wolf nearby was her mate, and he remained at the kill out of her sight. Was Explorer looking for an easy place to dig, or maybe a place with a microlandmark to allow her to remember the location, or something else?

I also wondered how many caches Explorer might have made on her stomach-load of meat. She spent from 0324 hrs till 0441 hrs on her caching mission and traveled at least 1.5 miles (2.4 kilometers) from the kill. With 22 pounds (10 kilograms) in her stomach (one full load), she could have made several caches. In northeastern Alaska, just after having fed on a carcass for 20 minutes, a wolf made six caches within 10 minutes all within 140 meters of the carcass (Magoun 1976). In August 1992, the Ellesmere wolves moved their three pups some 1.9 miles (3 kilometers) from the den, and when I finally found them, I watched while "Whitey dug up 3 caches, including 1 hare hind leg bone (more than a foot) in area of 100 sq ft."

How would these wolves find all their caches? Were they situating them in some mnemonic way so they could efficiently recover their contents?

Biologists have learned a great deal about food caching by various species, and the question of cache-content recovery has pervaded their studies of caching. Wolves are clearly "scatter hoarders" as opposed to "larder hoarders," which cache all their food in one place. The main problem scatter hoarders face is food recovery. With caches scattered far and wide around a wolf's expansive territory (some hundreds of square kilometers on Ellesmere), and possibly hundreds of food caches stored for days or much longer (as discussed later), it would take quite a memory to be able to find them.

If wolves could not recover a high proportion of what they cache, however, they would be wasting a considerable amount of their available food, not to mention the time it takes to cache it. Nevertheless, when I watched Explorer choose the two spots she chose to cache after making her kill in 1998, I could not discern any special characteristics about them, even though she spent several minutes looking around and choosing them.

When several years younger, this same wolf seemed to use food caching almost to taunt a packmate. On July 10, 1993:

> Explorer [yearling female] caught hare [leveret] at 1753 ~125′ from us and lay with it till 1804 before starting to eat it. . . . At 1813, Explorer took the hind quarters 175 m to w/in 30′ of WF [Whiteface, a yearling female] and nibbled them for 5 min, then took it 50′, dug hole and cached it [1823]. WF came to w/in 10′ and watched.
>
> WF then went to cache, and Explorer became very defensive—subordinately defensive while WF was very dominant. . . . This is opposite of their relationship near us and also opposite of what would be expected given that the hare was Explorer's. The posturing lasted 3 min. Then, both lay down near cache 6′ apart and slept.
>
> 1934—Explorer awoke and further buried cache for 2 min and lay back down near it when WF lifted her head and both slept.
>
> 1951 . . . suddenly saw WF pinning Explorer who was on her back for 2 min. . . . WF left, and Explorer dug up her cache and ate the hind quarter.

Explorer was a yearling at that time and perhaps had not learned yet to be more secretive while caching. This type of interaction with her pack mates may well be precisely the kind that teaches wolves this skill. Otherwise, they would lose much of what they cache since their packmates are so often nearby.

How much of their cached food do wolves usually retrieve? And when? Those are the important questions to answer in order to really understand the true value of caching, but, unfortunately, those also are questions that there are few ways of answering by most studies. A worker would have to be able to observe the wolves both caching and retrieving, which would require checking caches regularly for who knows how long?

It turns out that the Ellesmere study was able to provide at least some information about these questions. As mentioned, when I saw wolves caching something and was able to find the cache, I would mark it and then check it whenever I was in the area to see if it had been opened. Unfortunately, I could only be certain about the cache tenures within the same summers in which they were made because the natural ways in which I marked them did not necessarily hold up over the winter into my next summer study, or at least I could not always be certain they did.

Still, I did find that two caches were opened within a single day and one at 12 days, while others remained intact for periods of at least 9, 10, 16, and 29 days. The intact caches could have remained unopened for much longer periods, for the number of days I could check them was limited to my time on the island.

In fact, on July 3, 1994, I watched a wolf dig up a cache that I think was at least a year old, perhaps older (unpublished notes from Mech, cited in Peterson and Ciucci 2003): "Explorer [2-year-old nonbreeder] came down . . . and about 100′ SE of camp at base of hill, dug up a well-buried cache. It was a very old-looking (very dark and ragged and stringy leveret around 2–3 lb)—no doubt from last year. It was too large and old-looking to be from this year, and I have not seen 1 leveret yet this year. Explorer ate it and carried a small piece back up toward the den. Did *not* mark the cache."

It is also notable that Explorer approached the cache directly and began digging without seeming to sniff around or hunting for it in any way. I have seen wolves dig up other caches similarly, as though they knew exactly where they were. They seemed to either have remembered them for long periods, or perhaps as they passed them each time without opening them, they might have perceived cues that reminded them that the caches were there.

A good example occurred on July 12, 2004. I wrote: "she went straight to a cache and dug it up and ate several pieces (0020 hr). She then walked around 10 feet; tried to rebury a piece but digging too hard, so went to another place and buried it but then ate it. We checked the caches and found that the main one, which she opened, had contained about 14 inches of lower limb of m.o. yearling? and several chunks of well-aged (pasty) meat."

One of three primary theories of cache retrieval, in fact, is memory. The type of memory thought to be used for this purpose is "episodic memory," or memory associated with a specific event. Some types of birds such as jays and crows, as well as some rodents, are thought to use episodic memory to retrieve seeds, nuts, and other cached food. Other theories that may also be involved with episodic memory or perhaps used alone are "reforaging" and "search by rule." Reforaging is merely searching in familiar areas aided by cues such as odor, while searching by rule uses a simple rule or rules to guide the search, such as hiding food under certain-sized rocks (Breed and Moore 2012).

If Ellesmere's wolves used either of the last two approaches, that was not apparent to me. It appeared that those wolves I watched simply remembered where the caches were. Recent experiments with captive wolves showing that memory rather than scenting better explained their ability to find hidden food (Vetter et al. 2023) supported this observation.

Besides all the details about food-cache use, another aspect of caching, at least in canids, is related to some kind of symbolic property the caches appear to have. First discovered in red foxes by Henry (1977) and then in wolves and coyotes (Harrington 1981a, 1981b), some of these canids sometimes scent-mark caches after they have eaten their contents. With wolves, only individuals that urine-mark in other contexts mark caches. That is, although any wolf will open a cache and consume its contents, only mature wolves that mark areas with raised-leg urinations (RL urinations) or flexed-leg urinations (FL urinations) mark caches after they empty them (Harrington 1981a), behavior that remains an interesting challenge to understand.

My former graduate student Fred Harrington (1981a, 287), who later studied the marking of food caches in captive wolves, concluded: "Once an empty cache was marked it received little further attention, as opposed to caches that were empty but not urine-marked. These results suggest that urine-marking may enhance foraging efficiency in wolves by signaling that a site contains no more edible food despite the presence of lingering food odors."

A problem with this explanation, however, is that it does not account for why non-scent-marking wolves do not urine-mark their caches. Every wolf urinates. If the function of marking empty caches is to increase efficiency, why wouldn't every wolf tend to do it? I had been wondering this ever since Harrington did his study. The only other investigation of this phenomenon was also done with captive wolves (Townsend 1996), and the subject has never been studied in free-ranging wolves.

Thus, I tried to record information about caching every chance I got with the Ellesmere wolves. At times, I could only tell that a wolf was marking without discerning what kind of mark, but in 11 cases, I could determine that. The marking I recorded was done by four different females and four different males, including those featured in my publication (Mech 2006). Following are some of my general findings:

1. No wolf other than breeders or dominant individuals marked caches after feeding from them, similar to Harrington's (1981a) results.
2. Breeders sometimes marked their new caches but not always.
3. Males marked a significantly higher proportion (7 of 8) of opened caches than females (9 of 23).
4. Cache marking was by wolves' usual types of urine marks: squat urinations, standing urinations, RL urinations, or FL urinations or combinations, but females made all five squat urinations plus one FL urination, and males made all four RL urinations plus one standing urination. In one case, a female marked a cache with a squat urination, an FL urination, and a scratch (July 14, 1996).
5. In 1990, the first year that daughter Whitey became the breeder, both Whitey and Mom marked Mom's cache after Whitey opened it (August 1, 1990).
6. In 1991, the second year of Whitey being breeder, and Mom being the lowest ranking adult, I never saw Mom mark a cache, but the breeders did mark them.

None of my findings added any new information that allowed me to further judge whether the main function of marking empty caches is to signal that the cache is empty and thus save a wolf time in trying to determine that, although several years ago, I did tend to accept that explanation (Mech 2006). However, the male-versus-female difference in proportion of caches marked, the use of various kinds of marks, and the cache-marking behavior of Mom and Whitey all suggest that cache marking had more of a social function than an efficiency function, as Harrington (1981a) also recognized as a possibility.

The simplest explanation, and the one I now favor, is that wolves merely treat empty food caches as prominent locations where their marks are more likely to be perceived, similar to tree stumps, rocks, logs, and trail junctions (Peters and Mech 1975). Then, whatever functions any wolf scent-mark serves (see chapter 21), so too does the mark at the empty food cache.

It bears mentioning that, in addition to wolves scent-marking after emptying food caches, they also mark in relation to food remains or items that they usually eschew, like putrid food. (They do, however, sometimes eat putrid food [Mech and Breining 2020, 73].)

Although wolf marking of noncached, inedible food has rarely been documented, I summarized my Ellesmere observations on the subject as follows (Mech 2006, 468): "Breeding males and females used all four kinds of urine postures in marking uneaten food remains or prey odors including the following (1) Arctic Hare intestines; (2) chewed bones; (3) arctic hare stomach contents; (4) locations where regurgitated or other food had recently been eaten; (5) old kills; (6) the head of a recent muskox kill; (7) a putrid muskox carcass which they sometimes fed on but often passed up; (8) where another Wolf had tried but failed to catch a lemming; and (9) at caches of which they or another Wolf (including a pup) had just consumed the contents."

Here are some examples:

> 30 June 1991: While Whitey was running back and forth among the gut piles, she did a FLU near where one of the [hare] gut piles had lain. The remaining pile, probably mostly large intestines, was several feet away. Black mask then went over and RLU'd where Whitey had FLU'd.
>
> I threw Black mask a second [hare] stomach when Whitey was far away. He would not eat it, but RLU'd on it. He did eat [dry] dogfood and hare muscle. He also did not eat hare lungs and did not RLU on them, although this was just after he had RLU'd on the stomach.

> 7/5/95: Whitey SQU/FLU at the cache [my cache where I had stored 2 hares and recently dug them up] and then picked up a hare 1/2, that I had left out—took out of cache after 2 years, a few days ago—putrid—and carried it off and ate about 1/2 then left it. She returned to the other 1/2 and SQU/FLU.

: : **15** : :

Big Burly Beasts

Year-round, the Ellesmere wolves fed mainly on muskoxen, with arctic hares secondary, and the Fosheim Peninsula generally hosted ample numbers of both. Arctic-hare numbers varied considerably from year to year, sometimes reaching very low numbers (discussed later). For example, I received an email report that in 2020 one of the Eureka Weather Station folks had seen only a single hare between February 27 and the first week of June (Morgan Anderson, personal communication, July 6, 2020). However, muskoxen are much longer-lived, so their numbers would tend to be more stable. In 2017, their population on the Fosheim was estimated at 4,954 (95% confidence interval 3,461–7,091), a density of 768–1,572/1,000 miles^2 (300–614/1,000 kilometers^2) (Fredlund et al. 2019).

Wolves sometimes also killed Peary caribou and seals, but those prey provided only a small part of the wolves' overall diet (chapter 16).

Summers were probably the easiest periods for the wolves to succeed at their tasks because that is when both muskox calves and arctic-hare leverets, the wolves' main food by far, were available and most vulnerable. Muskox calves are born as early as mid-April on Ellesmere (Tener 1954).

Still, Ellesmere wolves don't always take just calves during summer. In 1998, when I could find no den nor pack of wolves to study but only a pair of wolves, I watched the two (Explorer and her mate) kill a cow muskox on July 8. Following verbatim from the *Canadian Field Naturalist* article that I and my companion and former graduate student Layne Adams published is the description of the attack and killing, which took place during just 5 minutes, an unusually short time:

> We spotted three Muskoxen about 500 m ahead in a valley at 0224 and immediately stopped and watched through 15X stabilized binoculars. The Wolves continued on toward the Muskoxen, and when about 100 m away, ran straight at them. The Muskoxen fled some 30 m and headed in a tight group up a steep slope, with the two largest animals (one a bull and the other presumably a bull) about half a body length ahead of the smallest, a cow. As the Muskoxen were running about a third of the way up the slope at 0226, the male Wolf grabbed the last one (a cow) by the rump and hung on, and the female lunged toward the head. The cow wheeled around, and the male lost his grip. Both Wolves focused their attacks on the head and neck of the Muskox, biting at her nose and neck, sometimes hanging on and sometimes losing grip. The Muskox kept pushing up with her lowered head and horns but did not use her hooves. After about 30 seconds of the focused attack, one Wolf gained a solid grip on the cow's nose and the other immediately attacked the side of her neck. repeatedly grabbing a new purchase. The cow appeared to struggle little once the wolves had gained solid grips on her.
>
> The two bulls had stopped about 15 m farther up the hill, and one of them suddenly charged down at the Wolves that were attacking the cow, sending one of the Wolves tumbling about 10 m down the hill. (We could not see whether contact was made, for the bull charged on the opposite side of the cow from us.) The bull hooked repeatedly at the remaining Wolf which eventually released its grip on the cow's nose. By now, the third Muskox had joined the other two, and they headed back up the hill with the cow tightly wedged between the 2 bulls. The Wolves quickly dashed back after the Muskox. Again one of the Wolves grabbed the rump of the cow, which wheeled to meet the wolf head on. The female then grabbed the cow by the nose, and the male by the side of the neck. The wolves kept their grips on the cow for about 30 seconds, and at 0231 the cow fell on a flat area of the hillside about 2/3 toward the top and stopped struggling. The Wolves continued to tear at her head and neck, but the Muskox did not move. (Mech and Adams 1999, 673)

Muskoxen might seem like one of the hardest prey animals for wolves to kill. Although of medium size, they are bulky and shaggy, with a huge blocky head sporting sharp, recurved horns; the creatures appear formidable. Muskox herds form defensive rings or lines, with each muskox facing outward and any calves behind or within the rings. "To confront such a façade would seem to a human to be foolhardy. So, too, to many wolves," wrote Mech, Smith, and MacNulty (2015, 129). It turned out that the muskox Explorer and her mate killed was in very poor condition, with only about 25% fat in her femur marrow, the last fat store in her body. That probably explained how quickly and easily the two wolves did her in.

Adult muskox weights vary considerably. Adult-male averages vary from 576 pounds (262 kilograms) on Devon Island (Hubert 1974) to 750 pounds (341 kilograms) on mainland Canada (Tener 1965), with females varying from 378 pounds (172 kilograms) (Lent 1978) to 460 pounds (209 kilograms) (Latour 1987; Schmidt et al. 2016). Newborn calves weigh 22–31 pounds (10–14 kilograms). Both sexes feature sharp, recurved horns, ever ready to hook attacking wolves. I once saw a wolf jawbone that had the tip of a muskox horn embedded in it. The calves are precocious enough to join their mother's herd when only a few days old. The herd affords the calves their protection via the adults trying to always stay between any attacking wolves and the calves.

Even the muskox's long, shaggy hair provides some defense against wolves. Guard hairs measure up to 24 inches (62 centimeters) long (Lent 1978). Wolves attempting to attack a muskox while both are in motion would find it hard to reach the body through all that shag and must also contend with mouthfuls of it in the process.

Nevertheless, under the right conditions, even a single wolf can kill a muskox (Gray 1970). It takes the right wolf, the right muskox, and the right conditions. In 1992, I found a place where the pair of wolves I was watching had run a bull into a river and killed it. From July 14, my notes tell the story:

> The wolves had attacked the [musk] ox on a river bar, but it apparently had then walked ~100′ down river and died on another bar. It had either half fallen into the river there or had been pulled there by the wolves feeding. My guess is that it had fallen there. The guts had been pulled out onto the gravel bar, but the rumen was still in the carcass. The lung cavity was open but lungs and heart were intact. I did not touch the carcass, but what I could see of one lung, I noticed no hydatid cysts.

> The carcass was on its right side with head and right side submerged. The left hind leg had been eaten and the bones gone. The meat along the left side of the vertebral column was gone and the whole left side of the abdomen behind the rib cage was gone.

It is mostly old bulls that live singly, and they make good targets for the wolves (Tener 1954).

Most muskoxen tend to live in groups, which afford them many of the same antipredator advantages as herds of other wolf prey (summarized by Nelson and Mech 1981). Muskox herds vary in composition. According to Gray (1987, 91): "Herds usually contain both males and females, but single-sex herds composed of all males and, occasionally, all females are also seen. Out of 94 single-sex herds seen between 1968 and 1978, only 5 contained females only. Herd size ranged from 2 to 8 for males, and 3 to 10 for females. Single-sex herds were seen between May and October, with the highest number in July, August, and September. Most herds were composed of adults, though subadult bulls were present in a few of the all-male herds." Solitary muskoxen are usually bulls that drift in and out of various herds."

Muskox-herd sizes are largest in winter (Hone 1934; Gray 1987; Heard 1992). On Bathurst Island, herds sometimes reached 60 in winter but usually varied from 5 to 15 in winter and 3 to 6 during summer. The largest herd reported was 100 on Queen Maud Gulf and on Melville Island (Gray 1987).

A study of muskox-herd size in relation to wolf densities across a variety of areas turned up some interesting relationships: "Variation in group size among areas was related to wolf densities in both seasons [summer and winter]. Seasonal differences in group sizes were related to winter snow cover probably because snow renders feeding sites less abundant, increasingly patchy, and more ephemeral. Group size appears to be a trade-off between increasing per capita benefits of decreased predation risks in larger groups at higher wolf densities, lower per capita benefits of group foraging in summer than in winter, and higher per capita crowding costs as group size increases" (Heard 1992, 190).

Given the relationship Heard (1992) found between muskox-herd size and wolf density, I would love to have used the herd size in his area to estimate his wolf density. However, the units that Heard used in his relationship were number of wolves seen per time during his flights. There was no way that I could relate that to my wolf observations from the ground, even though I did have a good sampling of muskox-herd size.

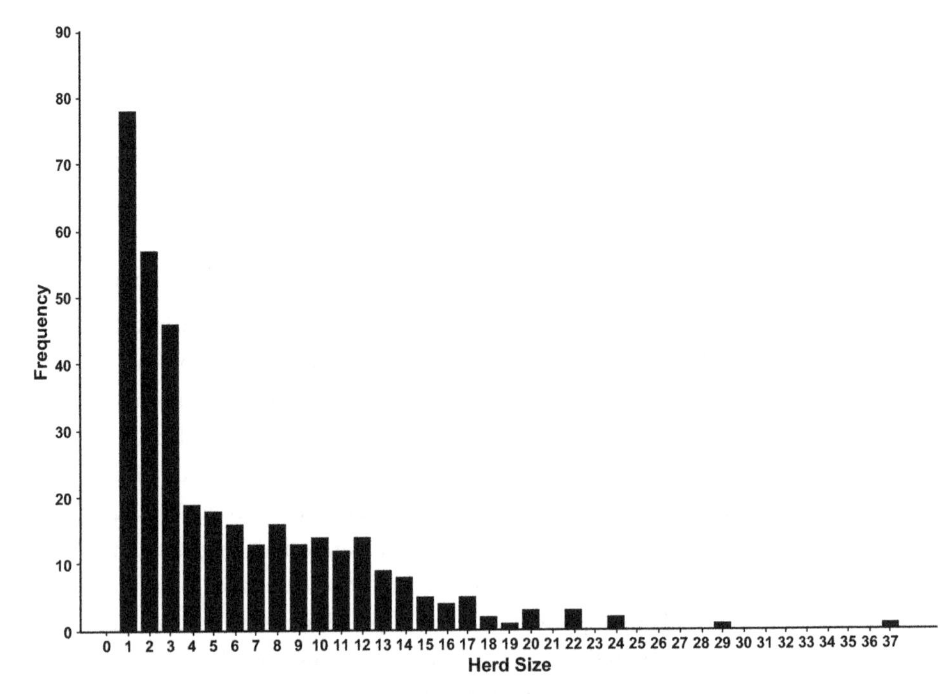

Figure 15.1. Size distribution of muskox herds that my colleagues and I observed on the Fosheim Peninsula, from the ground or air during 1986–2010, based on 360 observations of 2,116 muskoxen. The mean herd size was 5.8 and the median was 4.

Based on 360 observations of muskox herds in the Eureka wolf-study area during summers, my colleagues and/or I viewed herds of from a single individual muskox to 37 (fig. 15.1). I counted the largest herd (37) during my last minutes in the study area as I flew home on July 18, 2010, just north of an active wolf den south of Slidre Fiord; my partner Dean Cluff spotted 2 wolves at that den as we were leaving.

The composition of muskox herds, in general, is fluid, with individuals shifting from herd to herd at various times. Reynolds (1993, 88) wrote: "Muskox cows remain in social groups year-round as the benefits of group protection from predators and group feeding in winter outweighs the costs of competition for resources. These mixed-sex groups are flexible units which can aggregate or split into smaller units depending on local conditions. Muskox bulls have two different social strategies. Associating with mixed-sex or bull groups provides protection from predators during most of the year."

In the book *Wolves on the Hunt* (Mech, Smith, and MacNulty 2015, 129), I stated: "One of the benefits of herding is that it provides the group with the sensory abilities of each member. This advantage may be more important for muskoxen than for other prey because muskoxen senses do not seem to be very sharp." Neither I nor anyone else I know of has ever tested these senses, and little other information is available. However I have interacted with muskoxen many times on Ellesmere Island and near Aylmer Lake in the Northwest Territories. Overall, muskox eyesight seems poor. Usually muskoxen do not start grouping up until wolves approach to within 100–150 meters (109–164 yards), although I have seen them respond when wolves are as far away as 400 meters (436 yards). Muskoxen might not actually perceive approaching animals as wolves but might respond to any moving creature in a single stereotyped way. I once watched a herd group up upon spotting four white arctic hares running some 500 meters (545 yards) away. On the other hand, Gray (1987) observed a wolf surprise a prone bull by poking his nose into the bull's rump.

Muskox herds drift around over large areas. This behavior both helps herds forage on sparse vegetation that characterizes much of the muskoxen's northernmost range and helps minimize chances of wolves encountering them, a considerable benefit. Not much is known about muskox movement details, but the animals can trek long distances; some have traveled as far as 188 miles (300 kilometers) after being transplanted to new areas (Lent 1978).

On other Queen Elizabeth Islands, mean daily displacement of radio-collared muskoxen was less than 50 meters/day to up to 31+ miles (50+ kilometers)/day, based on distance between consecutive fixes that were usually four days apart. They can really move when they want to; but mostly they don't want to (Government of Nunavut, unpublished data). In their native environment, muskoxen may move in highly variable ways. On Ellesmere I saw that herds during summer often disappeared after just a few days in any given area. Individual herds seemed to drift around over areas as far apart as 31 miles (50 kilometers). Around Aylmer Lake in the Northwest Territories, muskoxen seemed to cover as much as 6 miles (10 kilometers) in just a few days.

Members of a herd graze for hours on end, being spread out as far as 50 meters from one another and the herd covering an area some 200–300 meters across. At the slightest disturbance, individuals stop grazing and become alert. If they confirm a disturbance, they quickly tighten up so tightly that their backs and rumps press against each other with no space among them except for calves huddling right behind them (Gray 1983).

Muskox herds roam the tundra nomadically, so wolves must often travel far and wide just to locate them. One measure of how much wolves have to travel to feed the pack comes from Brutus and his faithful GPS collar. Because the collar was programmed to only record two locations per day to save battery power and prolong the collar's life, the locations provide only an index of this travel. Nevertheless, they do also show how much ground wolves can cover during short periods. Brutus (and presumably his pack) in 12 hours once moved at least 47.5 miles (76 kilometers), measured by straight-line distance between two points 12 hours apart, and that was on December 19, when the area was in constant darkness (Mech and Cluff 2011).

Brutus's move that day was extreme but not surprising, given the amount of traveling I found when I did follow the wolves away from the den and will be seen in chapter 17. Even when wolves do find a herd, that is just the beginning of their real quest which is of course to catch and subdue one so they can eat it. Killing it happens in the process.

With most ungulate prey during summer, wolves tend to focus their predation on young-of-the-year, so I thought the same would be true with muskoxen. From summers 1986 through 2009 observing wolves on Ellesmere Island, I found that packs of 2–12 adult wolves killed seven calves, one yearling, and five adult muskoxen at distances of 1.8–20 miles (2.9–32.0 kilometers) from their current dens and pups (table 15.1). Of the 39 muskox kills made by wolves from July 2017 to June 2018 that Morgan and Dan MacNulty found by field examining wolf GPS location clusters, 15 (36%) were calves, including 3 neonates (Anderson et al 2019).

Given a possible bias against finding calves because of their fewer remains, these results accord with wolf predation on other species of prey. And given that calves usually comprise only 12–24% of muskox herds on the Fosheim, wolves seem to be selecting disproportionately for calves. Nevertheless, it is also clear that adult muskoxen also form an important part of the wolves' diet, even in July and thus probably at other times during summer.

Gray (1987, 129) described muskox interactions with wolves in detail based on 20 attacks he observed:

> Sometimes encounters do not progress beyond the wolves' initial approach to the herd, and the muskoxen do not group together in the defence formation. More often, once the approaching wolves

Table 15.1. Known muskoxen killed by wolves during July in the Eureka area, Ellesmere Island, Nunavut, Canada, 1986–2009

Year	Age	Sex	Number of adult wolves (number involved in kill)	Kilometers from den
1986[a]	3 calves		7 (7)	3.8–4.4
1987[a]	1 calf		7 (7)	32.0
1989	1 calf, 1 yearling		8 (6)	12.5
1989	1 calf		8 (6)	15.0
1990	1 calf		3 (?)	12.0
1992	1 adult	Male	2 (2)	2.9
1994[b]	1 adult	Female	4 (1–2)	2.9
1998[c]	1 adult	Female	2 (2)	–[d]
2009	1 adult	–	≥12 (?)	10.6
2009	1 adult	–	≥12 (?)	10.6

Source: Mech 2010.
[a] Mech 1988a.
[b] Well-worn teeth; femur marrow was fat.
[c] Mech and Adams 1999.
[d] Wolves did not den in 1998.

are perceived, an alarm snort is given and the herd groups together. The herd may stampede before forming a defence line or circle, and sometimes stampede after taking up a defence formation.

Once in a defence formation, most muskoxen, including cows, bulls and subadults, gland-rub when wolves approach to within fifty metres. The lead bull gland-rubs more vigorously, sometimes horning the ground, or rubbing his preorbital glands against the ground [thought to be a threat–Gray 1987].

In most interactions several different muskoxen (bulls, cows, or subadults) charge the circling wolf or wolves. Charges by single muskoxen facing one or more wolves may occur, with over twenty-three charges during a single interaction. Occasionally, part or all of the herd charges at the wolves together. Sometimes a group follows close behind as another muskox charges, preventing the wolves from cutting off the charging individual.

Muskox aggressiveness and the damage the shaggy creatures can deliver with hooves and horns usually deters wolves from pressing their attacks. Most often the combination of nomadism, grouping, and aggressiveness succeeds, and except for calves, old individuals or those debilitated in some way (Tener 1954, table 6), most muskoxen survive. This selective predation on the old, young, and debilitated is similar to the wolves' interactions with all their other hoofed prey species (Mech, Smith, and MacNulty 2015).

While watching wolves hunt muskoxen and trying to figure out what the predator's strategy was, I was confronted with the same bewildering antics I observed when conducting my first wolf study decades ago watching wolves hunting moose on Isle Royale: at times it seemed like the animals were hardly interested in some prey whereas a few hunts later they ended up killing their quarry. Somewhere, I wrote about all this to the effect of "the wolves often looked like amateurs" because at times when they approach prey it seems like the pack doesn't have its act together. Or perhaps the animals just aren't hungry. But then suddenly they make a kill and consume it feverishly.

July 4, 1989, was such a day. My companion and I had not yet found the den that year and had been searching for it for several days over a wide area. While heading back to our base camp at Eureka but still 11 miles (17 kilometers) away, we suddenly ran into the breeding male and five offspring.

At 1240 hrs the wolves left to the north, and I suddenly found that two had surrounded a grouped-up herd of eight adults and a calf but gave up in 1–2 minutes.

At 1320 hrs the wolves switched direction and headed southwest up a hill where I could see (a kilometer and a half away) several muskox herds. The canids quickly got to the nearest, about 800 meters away (nine adults and two calves), and at 1325 hrs they looked intently at the herd about 200 meters away, but the herd was already tightly grouped. Both wolves and muskoxen lay down at 1325–1405 hrs. Then the muskoxen grouped up, and the wolves left.

My field notes detail the story mentioned in passing in chapter 6:

> They continued W up the hill and disappeared and it took us awhile to catch up. When we did we came to a river gully draining the W Mountains and saw the wolves and various m.o. on a flat in the W side. There was a group of 4 cows and 3 calves grouped very tightly and about 200 m away a herd of 9 cows and 3 calves. In between, about 50 m from the smaller herd, stood a calf apparently confused about which way to

turn. The 6 wolves were speeding by the larger herd and heading straight to the calf. They all then grabbed the calf around the head as usual and pulled it down (1422 [hrs]).

. . . the wolves killed the calf within a few minutes. Then I noticed straight ahead of us—(the rest was taking place to our right and ahead)—what turned out to be a nose-wounded yrlg [yearling] . . . . they came back and worked on it, grabbing it mostly by the hind legs until they had it down. . . . They killed that animal within about 5 minutes also and began feeding on it. 1437 [hrs].

After killing the yrlg and calf and running between, the wolves noticed another single bull that we had seen 'hiding' in the creek gully, which was about 100′ wide and 15′ deep, run up the bank and try to get to the herd of 9 cows and 3 calves. The wolves headed straight toward him and tried to attack, but the bull whirled and charged them and within 30–60 sec, they gave up, and he then got into the herd.

This may not have been this bull's herd, however. There were 2 other adults elsewhere in the gully, which also came up to the larger herd. The wolves tried to attack them also but gave up quickly. However, within a few minutes, there was a fight between 2 bulls in the herd, with full head-banging 2–3 times and then the 2 squared off, but one ran to another part of the herd, while the other charged, so there was no clash.

The whole herd then began gradually to drift away from the wolves which were feeding about 150 m away. As the herd moved away, individuals grew antsy and began moving faster. Soon the whole herd was running toward the gully.

Instinctively the wolves stopped feeding and headed straight toward the herd, which fled across the gully and then stood on top of the E bank and fought off the wolves.

The last wolf stopped eating at the calf at 1552 [hrs]. At 1600 [hrs] 1 wolf came to the yrlg. where we were, ate a few bites and then headed SE across gully and disappeared at 1605 [hrs].

(My whole impression of this scene is that the terrain helped the wolves—perhaps they used the gully to get close to the m.o. and there must have been quite a bit of surprise and confusion since there were the calf stuck alone, the bull caught alone in the gully, and the other 2 adults in the gully—all away from both herds. Even the 2 groups themselves may have been a larger one that split.)

The wolves slept and rested after feeding.

Of note here is that two or three hours before killing the calf and yearling, the six wolves had approached two other herds including calves but quickly lost interest. Were the hunters sizing up those herds and somehow deciding it was not worth trying to attack them? If so, what clues gave them that impression? When they did kill the calf in the third herd, it was clear the animal was vulnerable because it was suddenly alone and confused. But then the wolves went for the yearling as well and took it down. Although already feeding on the two unlucky muskoxen, the wolves also perceived additional opportunities in the lone bull seeking the protection of the herd and then the entire herd once it began to flee.

I concluded that when wolves seem to forsake their quarry, it is almost certainly for a good reason. Reason that might not be apparent to a human observer but to highly experienced and savvy predators whose very lives and meals depend on them making the right decisions when confronting predation opportunities.

All this means that traveling far and wide and seeking as many predation opportunities as they can is the basic solution to the wolves' everyday challenge of securing enough prey to keep them going. It also implies that there will be far more failures than successes, as I had found elsewhere with other prey. With moose on Isle Royale, for example, a pack of about 15 wolves had a hunting-success rate in winter of about 7% (Mech 1966), or 6% when combined with similar observations by Peterson (1977).

With muskoxen in summer, I witnessed far more failures than successes too, although estimating the actual rate is confounded by just how the rate should be calculated. Should it be the number of muskoxen the wolves kill divided by the total number in the herd? The number of prey killed divided by the number of herds tested regardless of the size of the herd? What about the double kills like described above? One streamlined way to estimate the success rate is to merely take the number of successful hunts as a percent of the total number of hunts observed. With muskoxen, Gray (1987) observed three successful hunts out of 21, or 14%.

I decided to estimate a summer wolf-hunting-muskoxen success rate a couple of ways based on the hunting accounts I observed and described in my book *Wolves on the Hunt* (Mech, Smith, and MacNulty 2015): first the number of kills divided by the total number of muskoxen in the tested herds, resulting in a rate of 4% success. Second, the number of successes divided by the number of herds tested regardless of herd size and counting the double kill and triple kills each as one successful hunt, resulting in 21%, or one out of five hunts. The total number of kills I watched the wolves make included five calves, one yearling, and one adult (Mech, Smith, and MacNulty 2015).

These success rates were similar to the 1–56% hunting-success rates for wolves hunting moose, elk, caribou, Dall's sheep, white-tailed deer, and bison (Haber 1977; Carbyn, Oosenbrug, and Anions 1993; Nelson and Mech 1993; Mech et al. 1998; MacNulty 2002). (Wolves hunting moose in Sweden succeeded 45 to 64% of the time, but those estimates were based only on advanced portions of hunts during which the wolves were already chasing their quarry [Sand et al., 2006]). Thus, the Ellesmere wolf-muskox system seems to be functioning pretty much like most wolf-prey systems.

Although never one of the objectives of the Ellesmere wolf studies, the wolves' effect on the muskox population was always a logical question that lurked in the background. I did little to shed light on the answer, but my colleagues during 2014–2018 did take a good stab at that important topic, including a thorough muskox survey in 2017. That survey along with other data they had collected allowed them to calculate an estimated predation rate of 6–17% (Anderson et al. 2025).

Predation rates over about 10% with ungulates potentially lead to declines, and the above rate suggested that wolf predation could be a limiting factor for the muskox population in some years. However, it is also a rate that can be sustained by muskox recruitment, so declines due only to predation would not be expected. The occurrence of sporadic severe weather likely is a more important limiting factor for the muskox population. Like everything in ecology, it's probably complicated and a mix of top-down and bottom-up processes depending on the year and the scale.

Even my scattered counts of muskoxen over the years, whether from the ground or by air, always turned up reasonable numbers of the shaggy beasts. For example, in 2010 on one flight, Dean, Dan, and I counted more than 300 in just part of the general study area. Given the continued presence of the muskoxen as well as historical records reaching back to Tener's in 1954, it is clear that the wolf-muskox-hare system on the Foshiem Peninsula is an enduring one.

: : **16** : :

Hares, Caribou, and Seals

As with wolves everywhere, Ellesmere wolves feed on just about any other creature available. On Ellesmere, arctic hares stand out as the wolves' main secondary prey, and Peary caribou and seals are preyed on when and where they are available. However, in 1992, I watched a wolf bring home a fish, and at wolf GPS location cluster sites, Morgan and Dan found evidence of their having preyed on molting geese.

Still, besides muskoxen, arctic hares form the only other regular prey of Ellesmere wolves. They are distributed throughout the tundra zone of northern Canada, including all of the Queen Elizabeth Islands, the periphery of Greenland (except the eastern edge), and the island of Newfoundland. Adults weigh up to 15 pounds (6.8 kilograms; Best and Henry 1994), but on northern Ellesmere, 21 adult females averaged 8.6 pounds (3.9 kilograms) and five males, 10 pounds (4.5 kilograms; Caron-Carrier et al. 2022). Adults are white year-round on Ellesmere Island (though not elsewhere), undoubtedly reflecting the fact that it can snow there at any time of year. They sometimes burrow as far as 74 inches (188 centimeters) into the snow for shelter (Gray 1993).

Arctic hares live primarily on arctic willows, but they also feast on flowers such as the purple saxifrage that carpets the otherwise barren hillsides in summer (Bonnyman 1975). The creatures huddle down while nibbling on those spreads, but they can stand on tip toes and look around when startled. They also often run upright and drop to all fours when pressed. They can speed away at 38 miles (60 kilometers)/hour as Rolf Peterson measured on Ellesmere (Mech and Peterson 2003), similar to jackrabbits that have been timed at 35 and 45 miles (56 and 72 kilometers)/hour (Garland 1983) and about the same speed as wolves (Mech 1970).

Until recently, little was known about the general movements of arctic hares, for example, their home range sizes, dispersal distances, daily movements, and similar characteristics. However, the availability of satellite and GPS radio collars has prompted some studies that are now shedding new light on arctic-hare movements.

For example, we now know that arctic hares are extremely mobile. Not only do they travel far and wide, but they even seasonally migrate long distances. One adult female collared near Alert traveled at least 243 miles (388 kilometers) over 49 days, up to 19 miles (31 kilometers) between two daily locations and an average of 5 miles (8 kilometers)/day (Lai et al. 2022). This peripatetic hare might actually have been migrating, for several of her compatriots in the same area made similar synchronized movements. Several hares collared at Alert followed similar trajectories toward Lake Hazen covering around 125 miles (200 kilometers) minimum cumulative distance (Canon-Carrier et al. 2022).

Besides being so mobile and wide-ranging and able to run fast, arctic hares can zigzag at short intervals, much shorter than can a wolf, allowing hares a clear advantage while being chased by the much longer-legged wolves.

I watched a dramatic example of the hare's zigzag defense that almost succeeded while being pursued by an 11–13-year-old wolf on July 25, 1996: "The hare then jumped up, and the chase began. The wolf chased the hare for 6–7 minutes and almost caught it several times, but the hare's ability to make quick turns helped it elude the wolf since the wolf could not turn so sharply. The chase went back and forth, up and down gently sloping hills covering a distance with a maximum radius of 300 meters. At times, the hare was as far as 30 meters ahead of the wolf. Finally, at 0544 hr[s], the hare seemed to tire and slow down, and the wolf pounced on it" (Mech 1997b, 655).

The wolf carried the hare off and cached it. It weighed 54 ounces (1.45 kilograms). Two other wolf-caught and cached leverets that I weighed on July 16 that year were 42 ounces (1.2 kilograms) and 14 inches (35 centimeters) long and 46 ounces (1.32 kilograms) and 16 inches (41 centimeters) long.

Arctic-hare litters average 5.5 young (1–8) in late spring, and at least some hares might produce two litters (Best and Henry 1994). In June and July when leverets are young, most hares are dispersed across the tundra. The grayish-brown leverets match their surroundings. As they grow, they gradually become lighter until, when about three quarters grown by late summer, they become grayish white. The leverets are precocial, that is, born developed enough to hop away and fend for themselves while still needing to nurse from their mothers. The young are born in a fur- and dried-vegetation-lined depression, hollow, or crevice, but they leave the nest within a few days (Aniśkowicz et al. 1990). While still nursing, the leverets spread out and shelter next to rocks, in shallow depressions or in crevices. Each day, they congregate to nurse. Afterward, the female immediately departs, remaining increasingly farther from the site as the young grow.

Some leverets form nursing groups of up to 20 animals at about 3 weeks of age (Parker 1977). Congregating like this can help thwart wolves and other predators by maximizing their search time, spreading risk, and confusing the attackers (Hamilton 1971; Nelson and Mech 1981). On July 25, 1988, I surprised a single adult with 13 leverets that fled as a group. Three other leverets were within 30 meters of this group.

As the leverets begin reaching adult size by early winter, they group up with adults into herds, sometimes numbering in the hundreds (fig. 16.1). I observed hare herds of 40–50 in mid-July 2009, and Rolf Peterson while spelling me in late July 1989 counted a herd of about 200. I thought those herds might have been adult males because females should have been spread

Figure 16.1. As arctic-hare leverets begin reaching adult size by early winter, they join up with adults into herds, sometimes numbering in the hundreds. Photo courtesy of Eureka Weather Station/Rali LeCotey.

out at that time nursing young (Parker 1977). By autumn, most of the hares begin grouping in even larger herds for the winter. This grouping helps maximize wolf and other predator search time and improves survival, similar to the behavior of various ungulates in winter (Mech and Peterson 2003).

The value of grouping as a defense mechanism for leverets was quite apparent to me in 1996 when I watched the breeding pair hunting hares on July 25. Here from my field notes is the story:

> 0141 [hrs]—They [Whitey and her mate] were in the vicinity of the old m.o. carcass and hunted the gullies between the flats and Black Top Creek. They showed good teamwork, with each watching the other. Whitey tended to take her cues from the male. They often stopped on high points and scanned the area below for 30 seconds to 2 minutes.
>
> They headed N. along the gully above Black Top Creek, and at one point Whitey stalked a group of more than 8 leverets and suddenly ran toward them. I lost sight of them, but suddenly the male came charging in the opposite direction chasing a "cloud" of 16 light gray leverets. They all disappeared over the bank, and I saw individual hares running all over the creek flats and adjoining hills but lost track of the wolves (0200 [hrs]).
>
> We then caught intermittent views of both wolves till 0235 [hrs], and they did not act like they had caught any of the leverets. That is, they did not seem to be eating or carrying or caching any leverets and were not O.S. [out of sight] long enough for them to wolf down any of these, but not certain.

What came to mind while I was watching that "cloud" of light-gray fuzzballs streaking across the tundra ahead of the snowy-white male wolf was how confusing that chase would be as the targets started to spread out. Unfortunately, the chase took the wolf out of sight, so I could not watch the result. Still, it was clear a single target would have been much easier for a predator to keep track of than 16.

This well-known safety-in-numbers phenomenon is the most prominent feature of arctic-hare natural history that allows them to survive in the face of heavy wolf predation, but it is not the only one. Others are: (1) spacing out by the adult females and their offspring in summer; (2) precocial young ready to run and hide not long after birth; (3) the cryptic grayish coloration of young in summer, turning white in winter; (4) freezing in place by both leverets and adults; (5) high-speed and sharp zigzagging by young leverets older than about 4 weeks and by adults. Dean adds that hares chased by wolves tend to run up hill, perhaps giving them a running advantage.

Wolves, of course, know all this, and have learned various ways to succeed in overcoming these traits, at least some of the times. Wolves' sensory abilities are keen, as indicated by a time when I had been watching the wolves sleeping around their den during my first summer on Ellesmere. It was July 24, and I had finally got used to just sitting on my ATV and waiting for something, anything, to happen and break up the monotony. Maybe a hunting wolf would come home carrying a piece of prey. Or the pups would come out of the den and play.

Instead, I saw the breeding female, Mom, which appeared to have been sleeping soundly, suddenly stir and stare across the valley toward the top of the hill some 400 meters away. Up she rose and made a beeline toward that area. She headed northwest excitedly across a stream and up the slope toward an adult hare. The hare, being at the top of the hill, hesitated until Mom was around 50 meters from her to spring away.

Mom gave a very short chase and then started sniffing around where the hare had been. "Suddenly two or three small leverets (half the size of a snowshoe hare) sprang off in various directions. The wolf chased one downhill, and I found her eating the leveret at the bottom of the valley with several other wolves around her" (Mech, Smith, and MacNulty 2015, 149). Clearly this incident was an example of Mom's keen eyesight. She may also have understood that at least sometimes when she sees an adult hare, there may be vulnerable leverets around it, although perhaps she just spotted the young themselves.

With all the many hares that Ellesmere wolves attempt to capture almost daily, surely the wolves have learned many ways to overcome the hares' defenses, and no doubt the hares have evolved new defenses as part of the usual predator-prey arms race. The last half of a You Tube video illustrates a good example of the ongoing contest: https://www.youtube.com/watch?app=desktop&v=gGludGaPKag.

Evolutionary arms races like this between predator and prey have long been apparent to scientists, and Charles Darwin (1859, 77) featured them in his writings, although his studies preceded the term *arms race* itself: "the structure of every organic being is related, in the most essential yet often hidden manner, to that of all other organic beings, with which it comes into competition for food or residence, or from which it has to escape, or on which it preys."

Watching and studying these arms races was one of my favorite interests, and I had plenty of opportunities to observe the Ellesmere wolves as they preyed on hares. Because wolves need an average of about 7 pounds (3.25 kilograms) of food daily and eat a lot more when possible (Peterson and Ciucci 2003), it would take an adult hare each day or several leverets to fill one's stomach. That would mean that each wolf that counted on arctic hares for their meals would be matching their wits with those of the hares daily.

At times, meeting that challenge would be as easy as Mom realizing that the adult hare she had spied so many meters away might mean that some choice morsels would be surrounding that adult and be easily caught. Other times, it could mean a series of long chases as the leverets grew and developed. Eventually it might amount to many difficult pursuits ending in failure.

I got to see it all, at least during summer. Mech, Smith, and MacNulty (2015, 154–55) drew some basic conclusions:

> Wolves were successful at catching arctic hare leverets 60% of the time in July, with single wolves succeeding in 16 of 26 attempts and multiple wolves succeeding in 9 of 16 attempts. Anecdotally, it appeared that success rates dropped considerably as leverets grew, but the sample of observations was too small to allow for statistical testing.
>
> Following are some other wolf hare-hunting generalizations:
>
> 1. Much hare hunting is opportunistic—the wolves chase hares when they jump them while traveling or when the hares are resting or sleeping and the wolves spot them. . . . However, when wolves are resting or sleeping, they seem to be able to detect adult hares readily. The wolves must always be "keeping an eye out" even when seemingly asleep.
> 2. Wolves can see hares a long distance—probably as far as humans—and seem to be interested in the hares when up to 400–500 meters away. That is, when wolves see leverets as far away as this, they head to them.
> 3. The adult-male wolf seems to make use of lying-in-wait to ambush hares, especially when the yearlings are chasing them.
> 4. Yearlings seem faster at chasing hares than do the adults. (Mech wrote this in field notes in 1993, and MacNulty et al. [2009] tended to confirm it with wolves chasing elk).

5. Wolves can and do chase hares long distances, sometimes more than 800 meters.
6. When in an area with many leverets, wolves sometimes split up and lie not more than 200 meters apart and watch for hares.
7. When hunting hares, the breeding pair sometimes travel about 25 meters apart, probably to increase chances both of finding a leveret and catching one.

Because the manner in which wolves eat hares has not been described, I include the following excerpt from my July 19, 1994, field notes: "We watched [2-yr-old female] Explorer lying near [2-yr-old male] Grayback II while he ate an adult hare. He ate insides 1st, leaving stomach and guts, then down to hind legs and up to head, kind of peeling the meat out of the skin and eventually chewed through the skin holding front and back ends together, leaving 2 pieces. He looked like he was going to take the hind quarters to the den, but when Explorer [2-year-old female] came closer to him, he started eating the hind legs. He let Explorer poke through the guts and pick what she could but she left some of them. She ate the hind foot Grayback II [2-year-old male] left. (She had also eaten the hind foot the breeding male left.)."

It would be quite interesting to know just what proportion of the wolves' annual diet is comprised of arctic hares. About the only way to find out would be to collect a good year-round sample of wolf scats, which is almost impossible because of the constant darkness in winter. Tener (1954), who collected 85 wolf scats on both the summer and winter range of the local muskox herds, found hares in 83% of them. However, one cannot know how representative his sample was of the year-round diet.

Morgan and Dan MacNulty did come up with an estimate of the proportion of hares in the annual diet of one of the Ellesmere wolf packs. In July 2018, the two ground-checked GPS collar–based location clusters of two pack members during July 22, 2017, and June 19, 2018 (332 days). They found 39 muskox kills and 2 scavenges, and accounted for another likely kill at an inaccessible cluster, and calculated the amount of food those represented. They then figured out that the pack would have to consume the equivalent of 115–228 adult hares per wolf annually to make up the difference between that amount and the wolves' food requirements. The proportion of food from the hares would then form 34–52%

of the pack's annual food intake. If the two biologists missed a muskox carcass, or if the wolves' kill of those big beasts varied by a few, of course, that could change these figures considerably.

There is no question that during summer the Ellesmere wolves depend a great deal on hares, especially the leverets, and that is a critical period for wolves because the pups are growing and developing then. I lamented the fact that I had not thought during my first summer on Ellesmere to begin assessing the hare and muskox populations in some systematic way. Sure, my main concern then and for many summers thereafter, was to focus on wolf behavior, not wolf ecology.

It wasn't until 1998, when prey reproduction was so unsuccessful, that I realized I needed to start assessing prey in some fashion too. Recall that I never made it to the island in 1999 because of a back injury, but during my next trip I began my prey counts. And within a few years I realized the great value in doing so. During 2000 through 2006, I found a close relationship between number of adult wolves each summer and an index of the number of hares seen on a sample transect (fig. 16.2).

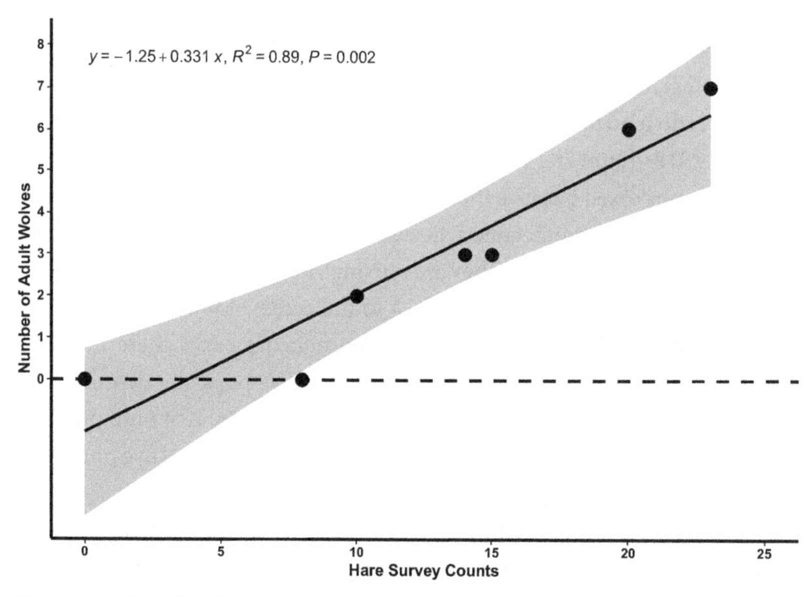

Figure 16.2. Relationship between adult-wolf numbers and arctic-hare index near Eureka, Ellesmere Island, Nunavut, Canada, 2000–2006 ($y = 0.331x - 1.25$, $R^2 = 0.89$, $p = 0.002$). Adapted from Mech 2007a.

A possible explanation for the relationship I found is that when hare numbers are high there would be less competition for them, so potential dispersing pack members can remain with the pack longer, so the pack is larger. When there are fewer hares, competition is greater, and maturing offspring that are often on the verge of dispersing tend to do so (Mech et al. 1998), so there are fewer adults.

In my 2007 article on the correlation I found between the number of adult wolves and the hare index, I closed with the following (2007a, 311): "The 2006 wolf population consisted of a breeding pair, five one- or two-year-olds, and five pups. Thus 5–10 pack members (42–83% of the population) could disperse before summer 2007, greatly reducing predation pressure on both muskoxen and hares. Future monitoring will help determine whether this reduction occurs or, if not, how the hare-muskox-wolf system shifts."

The system did shift. It turned out that the 2006 wolf pack left the study area over the next winter and began denning on the south side of Slidre Fiord, some 15 miles (24 kilometers) straight-line distance away. The main reason the pack shifted its den location may well have been because with the 2006 pack of seven adults and five pups, they took so many hares in their area that they then had to find a whole new area to den in where there was a better supply of hares. Although individual arctic hares can migrate long distances, resident adult females and males cover areas of only 0.27–0.60 miles^2 (69–155 hectares), respectively (Hearn et al. 1987).

Wolves do tend to den closer to their food supply (Ciucci and Mech 1992; Joly et al. 2018; Taton 2023). It would be difficult to confidently select an area with more muskoxen because those prey animals move nomadically, so their whereabouts are less predictable, whereas hares are local and thus predictable, at least during summer, when they are nursing young.

The new denning area that the pack found south of the fiord must have been quite favorable, for I documented reproducing packs there in 2008, 2009, and 2010. There were also two adults and five pups at a den north of the fiord but 14 miles (23 kilometers) from the 2006 den. Not until possibly 2011, but certainly in 2012, did wolves (at least three adults and five pups) use the original rock den they last occupied in 2006.

Once the wolves focused their hare hunting in their new area, the hare population in their former hunting grounds began greatly increasing. The index jumped considerably, from 0–23 during 2000–2006 (fig. 16.3) to 38–85 during 2007–2010. I was disappointed that I had not started a hare-index count much earlier in my studies, for it was becoming clear that the fate of

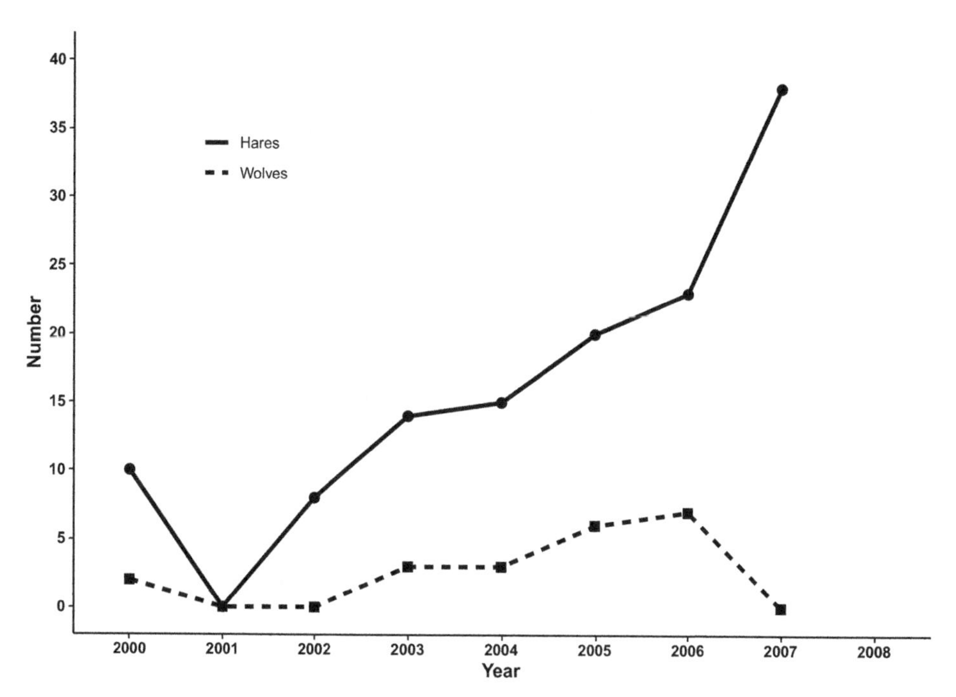

Figure 16.3. Total wolf numbers and arctic-hare index in the Eureka area (north of Slidre Fiord) of Ellesmere Island, Nunavut, Canada.

the wolves whose behavior I was studying was tied intimately to arctic hare numbers, at least during summer.

Besides feasting on arctic hares, the wolves I studied occasionally preyed on Peary caribou and seals. However, there were not enough Peary caribou on the Fosheim to constitute a very important part of their food supply. Peary caribou are the smallest caribou subspecies, with special adaptations for High Arctic living, including long, densely haired winter pelage, furry muzzle, short face, and short, broad hooves (Committee on the Status of Endangered Wildlife in Canada [COSEWIC 2004, vii]).

Peary caribou are continuously distributed across their range but are considered as four main populations—Banks Island (including northwestern Victoria Island), Prince of Wales and Somerset Islands, Boothia Peninsula, and Queen Elizabeth Islands (COSEWIC 2015).

The caribou that I observed on Ellesmere were few and far between, and the most I ever saw at once was 11 adults and three calves from the air (table 16.1). Aerial surveys by others in 2006, 2017, and 2019 in various parts of the

Table 16.1. Observations of caribou in the Eureka area of Ellesmere Island

Date	Caribou	Location	Remarks
June 26, 1989	1 adult	West side of Blacktop	Ground
June 29, 1989	two groups of 3 adults	North end of Blacktop	Helicopter
July 1, 1989	4 adults	West side of Blacktop	Ground
July 3, 1989	4 adults	Hill north of north end of Blacktop	Ground
July 7, 1989	5 adults, 1 calf; 3 adults	Blacktop	Helicopter
July 10, 1989	3 adults	West side of Blacktop	Ground
July 17, 1989	1 cow, 1 calf	West side of Blacktop; eating purple saxifrage	Ground
July 4, 1990	2 adults	Astrolab area	Ground
July 14, 1993	11 adults, 3 calves	Hills several kilometers north-west of Eureka	Ground
July 14, 1993	7 adults	Hills several kilometers north-west of Eureka; could be part of previous 11 but probably not	Ground
July 14, 1993	8 adults	East of Astrolab	Ground
July 7, 1994	1 adult	Flats northwest of Eureka	Groud
July 22, 1994	11 adults	Blacktop; seen byJay Hutchinson	Ground
July 1, 1995	8 adults, 2 calves	East side of Blacktop	Helicopter
June 26, 1996	1 adult	8 kilometers northwest of Eureka	Ground
July 4, 1996	1 adult	Eureka weather station	Ground
July 3, 2002	5 adults; 6 adults	Southeast side of Blacktop	Ground
July 4, 2007	6 adults	West side of Blacktop	Helicopter
July 4, 2007	7 adults	East side Blacktop	Helicopter
July 4, 2010	8 adults	North end of Blacktop	Seen by helicopter pilot

island also found low densities (Jenkins et al. 2011; Fredlund et al. 2019; Anderson et al. 2019). Those animals I saw mostly frequented the highest areas in the region, Blacktop Ridge northeast of the Eureka Weather Station and the Astrolab Peak west of Eureka. Every now and then I spotted one traveling the flats between the two highlands.

Nevertheless, a wolf did catch a caribou in June not long before I arrived in 1992, and a weather-station official happened to be in a position to videotape the attack. I described the attack from the video: "Excellent tape showing the wolf grabbing and holding the caribou by the head and tussling with it for several minutes. It actually had the caribou across the face and eye sockets and had completely torn out the top part of the skull, including the face from the ears down, and apparently had exposed the brain. All this while the caribou was still kicking." This was similar to what I had seen when the wolf was killing the calf muskox when we were filming *White Wolf*.

I also recorded wolves bringing a caribou leg to the pups in 1994. However, caribou were too sparse to have been a very important food source. In 2010, only one of the 17 sets of prey remains that Dean, Dan MacNulty, and I found by checking Brutus's GPS location clusters was that of a caribou (Cluff and Mech 2023). With field checks of 592 collared-wolf location clusters during 2014–2018, Morgan and Dan found no caribou remains.

With so few caribou on Ellesmere, I could not determine what ages or sexes the wolves took. However, my wolf and caribou studies in Denali Park, which I did concurrently with the Ellesmere work, came up with the same results as with wolves preying on most ungulates elsewhere. Denali wolves tended to end up with the youngest and oldest individuals and those with significant debilitations like malnourishment, parasites, and so on (Mech et al. 1998). Presumably that was the same on Ellesmere with the few caribou the wolves did kill. On Baffin Island, about 72% of the summer wolf scats contained remnants of caribou calves as opposed to 21% adults (Clark 1971).

An interesting possibility as to why so few caribou inhabit Ellesmere is because of a phenomenon known as "apparent competition" (Holt 1977), a term I find misleading and obfuscating. I would rather call it "subsidized predation." The idea is that when a predator has several types of prey and one type is plentiful, that type supports the predator, thus allowing the predator to also take rarer prey. That predation can then hold the population of the rarer prey down.

When such a relationship is viewed from the standpoint of the two prey species, "apparent competition" seems appropriate because the two populations seem to be in competition for survival against the predator. That is no doubt the way Holt (1977) saw the situation. Viewed from the predation perspective, however, "subsidized predation" seems more explanatory. Relating all this to the Ellesmere wolves, the question is, Are hares and muskoxen supporting the wolves and thus subsidizing them while they kill so many rare Peary caribou as to reduce their population?

I never collected enough information to allow any conclusion about this interesting relationship, but the possibility does remain. Several other workers have suggested apparent competition for caribou declines caused by wolves (Seip 1992; Hayes et al. 2003; Apps and McLellan 2006; Hervieux et al. 2014; Serrouya et al. 2017). Both Peary caribou and woodland caribou are sympatric with large populations of wolf primary prey (muskox or moose) that wolves can find more readily and subsist on. These caribou subspecies share a similar space-away, predator-avoidance strategy. That combination is the basis for apparent competition.

Canada listed Peary caribou as endangered in 2011 and in 2015 as threatened (COSEWIC 2015). Investigating the impact of wolf predation on caribou was the main reason that Morgan and colleagues were allowed to continue the Ellesmere wolf work in 2014. In 2007, a high density of caribou had been found on nearby Axel Heiberg Island (Jenkins et al. 2011). This finding would have allowed an interesting comparison of caribou ecology between Axel with a high density and neighboring Ellesmere with a very low density.

However, only six Peary caribou were seen on Axel when surveyed in 2019 (Mallory, Fredlund, and Campbell 2020) and only a few on the Fosheim Peninsula (Fredlund et al. 2019). Thus, further caribou studies had to be done elsewhere. A pilot project on Bathurst Island in 2019 showed promising results (Government of Nunavut, unpublished data).

Aside from caribou and hares, the only other notable occasional wolf prey were seals (*Phoca* spp.), although these sea mammals probably were not a very important food overall because they were only available for a short period in summer. During early summer when the ice of Slidre Fiord begins breaking up, seals pop up and lie on the ice near their holes, and wolves do take some. Seals are vulnerable when they come up on the ice to bear their young or to bask in the sun. In other areas where sea mammals are more available, wolves also readily prey on them (Griffin, Roffler, and Dymit 2023).

However, I only found evidence of such predation a couple of times. On July 14, 2004, Dean and I watched a pair of wolves we were following head out onto the ice floes to some seal remains about 80% eaten. The animals fed and rested between 1810 and 1848 hrs and then returned to shore. We later learned from military personnel that the wolves had killed the seal about 10 days before. Otherwise, the only evidence I found of the wolves eating seals during summer was on June 6, 1988, when one of the wolves and the pups were feeding on the hind half of a seal pup.

: : **17** : :

The Daily Hunt

The whole purpose of the adult wolves' travels away from the den was to procure food, not only for themselves but also for their pups. Except for the breeding female, other pack members tended to return to the den about once daily, and the breeding female twice as often, for she still had to nurse the pups. Most or all of these trips were hunting trips rather than strictly food deliveries, as evidenced by these wolves having only returned once daily, rather than more frequently. When wolves have a large food source, like a muskox carcass, they make several trips in a day.

When the Ellesmere adults left the den on a hunting trip, quite often they all departed together, usually leaving one individual behind, referred to as a "babysitter." This is not necessarily true of wolves everywhere. In my Superior National Forest study area, for example, individual pack members often hunted alone during summer (Demma and Mech 2009). Even on Ellesmere, individuals throughout the day sometimes forsake the hunting group and head off on their own. Thus, when members of the hunting party return for the day, they are usually alone.

Ellesmere wolves tended to leave dens on foraging trips more often during late evening and early morning, starting at about 2200 hrs, although they sometimes departed from the den area at all times of day, based on five of the Ellesmere summer, den-observation periods (Mech and Merrill 1998). The busy creatures left the den area least frequently on average from 1600 to 2200 hrs, which accords with the Anderson et al. (2019) findings. The lowest activity level the 10 wolves they collared in 2014–2018 occurred between

1800 and 2200 hrs, and movement rates of those wolves peaked from 0500 to 0800 hrs.

"Considerable ceremony often precedes the departure for the hunt. Usually there is a general get-together and much tail wagging." So wrote Adolph Murie (1944, 31) in an apt description of the Denali wolves as they prepared for a hunting trip. After the activity died down for a while, he concluded: "Then the vigorous action came to an end, and five muzzles pointed skyward. Their howling floated across the tundra. Then abruptly the assemblage broke up. The mother returned to the den to assume her vigil and four wolves trotted eastward into the dusk."

Such behavior also typified the activity of the Ellesmere wolves as they headed off each day too. Each time they left, I would love to have gone with them to see where they went, how far, and what they did. How long did they travel before resting? How often did they encounter prey and what was their response? Why did they split up, and what did each of them do when they were on their own? There would always be so much to learn during every trip they took. Nevertheless, there were the pups themselves to watch and all that went on around the den between the pups and the wolf (or sometimes wolves) that stayed behind and minded them so often. And then there were the returns of the hunters to the pups, which were also quite exciting and informative.

And I did follow the adults on their hunts at times, so I was able to get some of the kinds of information I wanted during their forays. For one thing, I sometimes timed them for different distances to see how fast they traveled. My ATV had an odometer, so as long as I kept track of the start and stop times and odometer readings for periods of travel, I could estimate the wolves' speed. I did this at various times for single wolves, a pair, and a pack of five (Mech 1994a). Their mean travel speed was measured on barren ground at 5.4 miles (8.7 kilometers)/hour during regular travel and 6.0 miles (10.0 kilometers)/hour when returning to a den (table 17.1).

The regular speed of these wolves at their normal pace was about the same as I had found decades earlier with wolves on Isle Royale crossing a frozen lake (8.0 kilometers/hour), which I was able to measure from an aircraft (Mech 1966). Then there was also Brutus, which Dean and I GPS-collared in 2009. Although the average straight-line distance between his locations 12 hours apart was 6.9 miles (11 kilometers), at least once he traveled 47.5 miles (76 kilometers) in 12 hours, an apparent record (Mech and Cluff 2011). That was a minimum speed of 6.3 kilometers/hour. Because Brutus probably didn't

Table 17.1. Travel speeds of wolves, Eureka area, Ellesmere Island, Nunavut, Canada

Date	Distance followed (km)	Animal	Speed (km/h)	Remarks
June 27, 1989	1.7	Breeding female	12.0	Returning to den
June 27, 1989	4.0	Breeding female	9.6	Returning to den
August 2, 1989	1.8	Yearling male	13.1	Returning to den
July 11, 1993	1.3	Pack of five	9.6	Regular travel
July 11, 1993	1.2	Pack of five	7.8	Regular travel
July 11, 1993	0.9	Pack of five	6.6	Regular travel
July 13, 1993	1.2	Pack of five	9.0	Regular travel
July 13, 1993	1.8	Pack of five	9.6	Regular travel
July 16, 1993	1.4	Breeding pair	9.3	Regular travel
July 16, 1993	4.0	Breeding pair	9.6	Regular travel
July 20, 1993	4.9	Breeding pair	7.9	Regular travel

Source: Mech 1994b.

travel directly in a straight line during the whole 12 hours, any deviation off that line would have meant traveling even faster.

I had followed wolves from an aircraft during three winters on Isle Royale and for parts of many other winters in Minnesota's Superior National Forest and kept track of how far and wide they traveled. Once I monitored their movements for 31 consecutive days on Isle Royale and found they covered 277 miles (443 kilometers), or about 8.8 miles (14 kilometers)/day (Mech 1966). However, for 22 of those days they were feeding and resting around kills. Thus, for the days they were traveling, they covered about 31 miles (50 kilometers)/day. Nevertheless, that was all in winter when they were nomadic within their large territory and thus had nowhere else in particular to be.

How far would the Ellesmere canids travel when they were committed to returning each day to a den to feed their pups? I got a pretty good answer fairly early on, during my second summer with the pack. One day, I followed the wolves as they traveled about 24 miles (39) kilometers to where they killed the muskox calf (Mech 1988a) featured in the National Geographic documentary *White Wolf* (mentioned earlier). Then they had to return to the den too, a straight-line distance of about 20 miles (32 kilometers), so they traveled about 44 miles (70 kilometers) that day. Years later, after the pack moved to a den south of Slidre Fiord in about 2008, for several days they traveled from that den to the Eureka area itself, a round-trip of 25 miles (40.2) kilometers, plus whatever additional distance they covered while actually hunting (Mech and Cluff 2009).

"The wolf is kept fed by his feet" according to an old Russian proverb (Mech 1970), and whether it be during most of the year when wolves are nomadic within their territories, or when restricted daily to returning to feed the pups, that rings true. The reason is simple: to find enough prey that they can catch and kill regularly, wolves need to travel far and wide.

A question that plagued me about wolves traveling each day from the den area is what might be going through their minds when they leave, or at least through the mind of the wolf that leads the pack away from the den. As for which wolf actually leads the pack away, I found that this is not a simple question either (chapter 18).

Regardless of whether or when the breeding male tends to feed the breeding female, or which wolf tends to take the lead as the pack heads off each day from the den to procure food, there is still the question of what goes through that wolf's brain as the animal leaves? Does that wolf have a destination "in mind"? For example, Today we'll head over to Eastwind Lake and see if there are any muskoxen around there.

Of course, there is no way of knowing. It is true that as the wolves left the den area on a hunt, they did not merely wander around as though they were traveling randomly. No; they usually beelined it away, looking as though they did have a destination. And each day they seemed to take different routes away and cover different areas, a practice known as rotational use.

In studying the very first wolf ever GPS-collared, I and my colleagues in Alaska's Denali National Park found that it tended to head away from the den into different directions each day and concluded that the value of doing so might be to increase the chances of surprising more prey (Mech et al. 1998). Prey harassed by predators tend to show "behavioral depression," including increased alertness (Charnov 1976), so it would be adaptive for wolves to temporarily avoid areas they had recently used (Jędrzejewski et al. 2001).

Studies elsewhere confirmed that rotational use of their territory is not unusual for wolves. That is, on a daily basis, wolves tend to avoid areas used previously. In Poland, packs tended to use new areas each day primarily during fall and winter, covering the whole territory in about 6 days (Jędrzejewski et al. 2001). In the Superior National Forest of Minnesota during the denning season, the mean daily overlap of areas used daily was only 22% (Demma and Mech 2009).

Wolves' rotational use of their territory implies that they do possess some sort of mental processing and that they have in mind a kind of spatial representation of their surroundings. For many decades, scientists have thought that most mammals host some kind of mental maps, known as cognitive maps, and one of my early graduate students, Roger Peters, published evidence that the wolves he and I were studying in the Superior National Forest of northeastern Minnesota harbored such maps (Peters 1979).

Peters followed wolves through the snow day after day and noticed them shortcutting to other of their trails, even though they had not ever used the shortcut trail. They would also shortcut a curve in a trail, but Peters only considered a shortcut to be evidence of a mental map if the trail they shortcut was at least 0.6 mile (a kilometer) long. Shortcuts could not be explained by rote learning, and Peters ruled out long-distance orienting because that involves long straight line travel. Use of a cognitive map was the only other possible explanation. That map guided them across new ground that made their journey shorter.

Thus, Roger and I used to discuss such topics. That came to my mind one day when I was in the back seat of a light aircraft with a semidrugged wolf on my lap flying the wolf back to its territory in the Superior National Forest. I had radio-tagged the animal long before but had recently rescued it from languishing illegally in some trapper's fox trap, held it in a cage and fed it for a few days, and was now on my way to releasing it back home (Mech, Seal, and Arthur 1984).

The wolf, bound and gagged, began to awaken a bit and looked out the window at frozen Quadga Lake below. The thought that immediately went through my head was whether the wolf was thinking, as a human might when seeing the lake from the air for the first time: "Oh, gee; there's Quadga Lake!"

Given all that is known about honeybees, ants, and other insects navigating long distances, it is not all that surprising that wolves would have cognitive maps with which to guide them in their long travels. As it turns out, science even knows where in the brain those maps reside, the hippocampus. There, cells discovered fairly recently, known as grid cells (Leutgeb et al. 2005), form the basis for spatial representation in the hippocampus (Bush et al. 2015).

For a few years, the idea of animals possessing mental maps fell out of favor. A *Scientific American* article, "Rats, Bees, and Brains: The Death of the 'Cognitive Map'" (Goldman 2011), seemed to have killed it. However, in the beautiful way science advances, current thinking is that there is good evidence that at least some mammals do have such maps (Harten et al. 2020; de Guinea et al. 2021).

It's a good thing they do have such maps, because with the wolves' widely ranging around their territory to find enough prey to catch and kill, they need to travel efficiently. Otherwise, they would not have enough energy to catch their prey once they even found it.

Peters (1979, 126) put it this way:

> An efficient means of remembering location is particularly useful to social carnivores, for several reasons. The large animals they eat are less densely distributed than the small animals and vegetation that nourish other creatures. Social carnivores like wolves must therefore travel widely in order to encounter and test as many potential prey as possible. Furthermore, the distribution of their prey is often clumped . . . which places a premium on knowing where to look. And, finally, social carnivores must often not only find prey, but also find their way back to a den or rendezvous site in order to feed their young.

So the Ellesmere wolves probably make use of mental maps that are spread over the grid cells in their brains to help them find their way around their territories as they search for prey they can kill. Perhaps they recall where especially defensive individual prey or groups hang out, so they do not waste their time on them; or conversely, certain areas where the terrain facilitates their hunts ("terrain traps"). With the horrendous problems wolves face each day to find vulnerable prey, a mental map would surely be beneficial.

And the pack leaders do seem to have specific destinations in mind when they start out, usually in early evening (Mech and Merrill 1998). They travel tirelessly at about 5.4 miles (8.7 kilometers)/hour searching for prey they can kill. Then, knowing where everything important is, they can efficiently bring the spoils from the kills back to the breeding female and/or the pups.

I would love to have followed the wolves on their daily hunts, but most of the time I prioritized sticking around the den to record the behavior of whichever wolves remained there each day. In a study of Yellowstone elk, colleagues and I had analyzed wolf travel during 2004–2012 and detected that they encountered elk 0.79–0.94 times per hour (Martin et al. 2018). The park's elk density was relatively high as wolf-prey densities go (Mech and Barber-Meyer 2015). Thus, Ellesmere wolves would probably have experienced a much lower rate of encountering muskoxen.

Both watching the wolves' behavior around the den and following the adults on their hunts would have required two teams, which would have been prohibitively expensive. However, during the first couple of years when originally trying to figure out the best approach to gain the best data, I did follow the pack at times to learn where they went and the details about what they did on their travels. One of the most interesting days was July 31, 1987, when the photographer and I tracked the pack for most of the day when they killed a muskox calf (Mech, Smith, and MacNulty 2015).

At 0930 hrs we spotted all six adults still heading north several kilometers from the den and about 2 miles (3 kilometers) west of us. We headed there and northward on top of a hill and saw the pack about 4 miles (6.5 kilometers) north, heading toward a herd of about eight muskoxen. We headed there, but by the time we arrived, the wolves had left. We continued north and began heading northeast 2 miles (3 kilometers) farther and crossed a river with steep banks only a few kilometers south of Greely Fiord.

The pack had found a group of two large and one medium muskoxen standing their ground. The wolves lay down near their prey about 12 meters away but were not interested in them until their targets started running. Each time the burly beasts fled, the wolves chased. But usually the muskoxen stopped within 25 meters, so most chases were short. I thought it possible that the muskoxen were running from the ATVs and that otherwise they would have just stood their ground until the wolves left. I decided to check this by remaining farther away sometime and watching from more of a distance.

The three prey ended up standing their ground in a gully 5 or 6 meters deep. My partner and I ate lunch while the wolves slept for about 45 minutes. One large and the medium muskoxen eventually left the gully, while the other remained in it with its rump up tight against the bank. (During one of the chases, the smaller animal got separated from the other two, and one of the wolves actually nipped at its hind leg but didn't seem to get anything but hair.)

At 1430 the wolves left and headed east to northeast up a hill and seemed quite intent during their travels. We stayed with them. In one place the pack did much scent-marking on some rock formations at a prominent point. Two males RL urinated at once, and a female squat urinated or defecated, and it appeared that all wolves marked, some two to three times although I wasn't sure of that. They headed on, and about 800 meters further they found a single large muskox which they chased briefly and gave up on. They did not even seem interested in it. Did they know this individual from a previous encounter?

Several hundred meters beyond that, heading eastward near a long, small lake, the pack appeared to sense a herd of about 14 muskoxen with at least two calves from about 300–400 meters and headed for them. The wolves were coming down a slope eastward and curving around to the north down a rocky slope leading to a wet flats with green vegetation along the west end of the lake. They were all spread out, some being 150 meters from the others.

The wolves spied the herd from about 250 meters and began stalking them slowly, deliberately and stiffly. Although the wolves were strung out, even the last animals 100 meters behind were just as still. When the lead wolf—which I thought was Mid-Back (Mech 1988a)—was within 150 meters, it charged down the hill, and all the wolves followed. Immediately the muskoxen grouped up, and the wolves swirled around them for about 1–2 minutes, and then seemed to give up and head beyond.

Suddenly the herd charged up the hill just east of where the wolves had come from. However, as soon as the herd started running, the wolves took chase (fig. 17.1) and the herd rushed right toward the photographer and went by him. He continued filming and said the wolves hit one calf as they went by him, and he filmed it. The herd split, part running south and the other, east. The wolves chased those heading east.

My notes best described the situation:

> I was trying to stay out of the way of his filming, by remaining below the hillside the m.o. were heading up. However, when they had gone by, I headed up to watch the chase and found that a calf had detached from the herd and was hanging around the photographer and his 4-wheeler. Most of the wolves were giving up from chasing the E group and heading N after the N group, but 1 male lagged and spotted the lone calf.
>
> About 60′ from, me, he grabbed the calf by the nose and had a tug of war. I had no camera, so sat and watched. The calf bawled, and one female wolf came shooting straight and grabbed its left ear. More wolves then arrived, and they all grabbed the calf around the head and had a tug of war.

Figure 17.1. As soon as a muskox herd starts running, the wolves take chase.

> At one point, the calf shook all the wolves and headed W toward the photographer who had moved closer. Its face was torn out and bloody. The wolves caught it after a few yds and continued tugging on it till it went down, about 10 min.—estimated—from the time the first wolf grabbed it.

By this time, the photographer was nearer to me and shooting west toward the kill. He was changing film magazines when the calf actually went down. When we both drove over to within 20–30 meters of the wolves and their fresh food, he continued filming them at their feast. At first when he dismounted and set up his tripod, the wolves left the carcass, and several headed down toward the lake 150 meters away.

> After a few minutes, the alpha male came to the carcass and fed, and so did the other wolves. Various wolves fed for various times and came and went and at least 3 cached food usually several hundred meters away.
>
> The alpha male again did much scratching, and one wolf sniffed the scratched ground. The alpha male also squat urinated [*sic*] and scratched the area.

> At one point, one male and Shaggy approached alpha male at the carcass and were very submissive. The male pawed alpha male's head from a half-sit, half crouch, and Shaggy crouched even lower, and they licked alpha male's face very submissively like pups while he stood rigidly. Eventually they ate while he did also.
>
> We changed positions and filmed from the west, and I sneaked on all fours toward the carcass while the alpha male was feeding and got to within 50'. One of the other wolves left while I took notes.
>
> Eventually as the carcass got lighter the wolves began to drag it downhill away from us. Two wolves left the area about 1830 hrs. The alpha male, 1 other and Shaggy remained. When we left, at 1855 hrs., the carcass was about 2/3 eaten or gone. I estimated its intact weight at about 125 lbs.

Biologist and photographer, thrilled, excited, and overwhelmed, we headed back toward camp in almost a straight line, refueling at a fuel cache partway back. We arrived at the den at 2205 [hrs], and one wolf (almost certainly Mid-Back) was there, along with Mom. (We had seen two near Eastwind Lake while refueling about 2030 [hrs].) The odometer distance from the kill to den was 24 miles (39 km) and map distance, 20 miles (32 km).

The last three wolves arrived at the rendezvous site near the den about 2245 [hrs]. Mid-Back, which was already there (or possibly it was Shaggy), took three pups up the hill to the south and out of site. She regurgitated to one. There was much howling for about an hour, and the wolves were quite scattered. This killing of the muskox calf described above was the one mentioned earlier that was featured in the National Geographic Explorer film *White Wolf* (https://www.youtube.com/watch?v=pb6Rke7jiTcI). I referred to the breeding male in the previous notes as "Alpha Male." Those notes were written before I had published my article "Alpha Status, Dominance, and Division of Labor in Wolf Packs" (Mech 1999), in which I suggested abandoning the use of *alpha* for wolves (chapter 7).

The wolves were gone from the den for almost 13 hours to a point some 20–24 miles (32–39 kilometers) away, with a travel distance more like 25–30 miles (40–48 kilometers) one-way. Their long travels paid off, however, by their having found the herd with calves, and they cashed in royally by nailing one. During that whole operation, it seemed like the wolves' basic hunting strategy was just to charge the muskox herd, separate out a couple of calves, and attack them.

Sometimes, however, it does seem like these predators are more strategic. In 2006, while I was watching seven adults and five pups around the original (1986) den, I observed a couple of interesting hunts right from the den area. During them, the wolves did seem to demonstrate some more complex strategy (Mech 2007b).

In one case, on July 2, I spotted a herd of seven adult muskoxen and three calves. All seven wolves headed north, jumped two hares but did not chase them (which was unusual), and continued on to the northeast. I could see the herd from where the wolves and I had been, so presumably the wolves had also seen the muskoxen from the den area. The wolves disappeared until about 2250 hrs; they reappeared about 200 meters from the muskoxen and headed toward them up a shallow valley, slowly stalking them.

At least four other wolves were watching intently from a ridge of rock piles approximately 400 meters from the muskoxen. Suddenly the muskoxen ran to each other, two to three that were lying down arose, and all grouped up. Then all the wolves, both waiters and stalkers, rushed to the herd, their movement apparently triggered by the muskoxen fleeing. The wolves milled around the herd for about a minute, then left and continued north.

Waiters and stalkers. Had the wolves deliberately split up and decided to ambush the herd? That's what it looked like to me. The ploy hadn't worked, but under other circumstances it might have. Wolves have a lot of time. There will be other herds to test. As I had long been fond of saying after watching Isle Royale's wolves test so many moose before they finally found one they could kill: "What else have they got to do?"

In fact, I even got another chance to watch some fancy wolf hunting antics from the den area a few days later. On July 6, I could see about 13 adult muskoxen and 7 calves coming into view about a mile (1.6 kilometers) northeast of the den. At 1628 hrs, male subordinate Wolf A, while on a ridge just east of the den, stared intently toward the muskoxen for 1 to 2 minutes. He then headed to another subadult wolf (Wolf B) of unknown identity, which was lying about 20 meters away below the ridge chewing on an object, and "nosed" that wolf. They were too far away for me to hear whether either made any sound, but Wolf B immediately abandoned the object, went to where Wolf A had stared toward the muskoxen, and also stared toward them. It appeared that Wolf A had communicated with Wolf B, motivating Wolf B to look toward the muskoxen.

At 2025 hrs two wolves, Wolf A and a wolf that was either Wolf B or one of two other possible subadults, had gone down the den valley and then up onto the flats to within 300 meters south of the muskoxen. The wolves had moved around 200 meters on the flats toward the muskoxen when they stopped, stared toward the herd for a few minutes, and then backtracked about 50 meters. They headed east about 100 meters and then back north about 50 meters, moving toward the muskoxen along a parallel to their original route.

They then lay down out of sight, about 100 meters from a green, wet sedge meadow about 15 meters wide and 40 meters long, still about 300 meters from the herd. (One adult muskox stood around 200 meters southeast of them, possibly having passed from the herd through the area where the wolves now stood before the wolves' arrival.)

The two wolves lay hidden around 200 meters southeast of the muskoxen at 2034 hrs; the muskoxen grouped up loosely, but by 2042 hrs had resumed feeding.

At 2104 hrs three adult muskoxen meandered down the meadow near the wolves. When the first was within around 30 meters of the two secreted canids, these patient creatures left, sneaking around 30 meters south down a trench. They circled west and lay down in a hidden spot around 15–30 meters from the meadow, in rough, uneven terrain with lots of trenches and small hillocks (1 meter high). The muskoxen did not appear to sense the wolves and continued southeast. The herd, some 200 meters away, lay down. I continued to watch from the den with 15× stabilized binoculars. The muskoxen and wolves remained where they were for three hours. At least one of the wolves intermittently lifted its head and apparently was watching the muskoxen. From my notes:

> 2330 hours—The muskoxen began to arise. 2337 hours—The muskoxen grouped momentarily (30 seconds) and then opened the group again and began drifting southeast along the same general route as the first three adults.
>
> 2342 hours—The muskoxen got to within 100 meters of the wolves, which then circled east and then north up the trench and charged the muskoxen from about 50 meters. The muskoxen grouped, and the wolves milled around about 30 meters away for 1 to 2 minutes, then left and lay down around 60 meters from herd. After around two minutes, the muskoxen continued drifting southeast. The wolves arose and muskoxen grouped for one minute and traveled on. The two wolves lay back down and slept.

This attempt had been a bust, like most of them. But it did give me a number of insights into some pretty interesting hunting behavior. It certainly looked like the wolf that had first spotted the muskoxen somehow prompted the other to view them too and that both waited hours until they were ready to hunt. Then they sneaked toward the herd, positioned themselves, and adjusted their positions to best ambush the herd in an area where the muskoxen eventually would feed, and waited there until the herd came by.

These wolf-hunting-behavior observations seemed to indicate that at least during some hunts, wolves are capable of such higher-order mental processing as understanding, foresight, and planning (Mech 2007b). However, some scientists have attempted to explain these observations of wolves hunting as the wolves simply following two rules that require no such higher-order processing: "(1) move towards the prey until a minimum safe distance to the prey is reached, and (2) when close enough to the prey, move away from the other wolves who are close to the safe distance to the prey" (Muro et al. 2011, 192).

Although following these rules might explain many wolf-hunting attempts (Mech, Smith, and MacNulty 2015), it is hard to see how they address these observations. In the earlier case, the wolves split up before reaching a minimum safe distance from the muskoxen, with one part of the pack lying in wait while the others stalked the prey. It is true that once the muskoxen grouped up and both waiting wolves and stalkers approached, only then did the canids stop at a minimum safe distance. But their preceding behavior did not seem to fit the Muro et al. (2011) model.

In the second case, the pair of wolves seemed to size up the situation from a long distance, then approached not to a minimum safe distance but to a position to lie in wait and then even changed their ambushing position. Only when the muskoxen moved closer to the wolves, did the wolves charge to within a minimum safe distance. Again, their preceding behavior failed to fit the model.

Neither of these attempts succeeded, but sooner or later another would. Then, like during the 1987 hunt that I watched up close, after the wolves scored, they would gorge, maybe cache some fresh bloody goodies somewhere for later picking, and then head home to feed the pups. They would literally "bring home the bacon." Day after day, that is the job of the adults, and if they do it well, their pups survive, grow, and develop, and the adults' own genes get passed on.

At least during the periods each summer when I was watching all this, the wolves did do their jobs well, for not one pup died during those times. Of course, some pups could have perished before I arrived, and certainly some died after I left each summer.

:: **18** ::

Divisions of Labor

From an evolutionary standpoint, every living thing's primary function is to ensure maximizing the future of its DNA or passing its genes on to the next generations (Dawkins 1989; Hamilton 1964). In this respect, almost any wolf behavior will serve that function either by helping survival or reproduction. Because of sex differences, however, some male and female wolf behavior will differ. This is the case even though, contrary to most mammals, both wolf sexes are critical members of the wolf's basic social group, the pack. In fact, the pack is the social group best suited to foster the survival and reproduction of both male and female wolves.

For the basic survival of the pack, each sex contributes, sometimes similarly as in protecting pups and sometimes in its own way (J. Smith et al. 2022). Some of these different gender roles are well known. For example, males are critical to the protection of the pack against other packs (Cassidy et al. 2017), whereas we biologists are still trying to figure out other male roles as discussed later.

When it comes to promoting their own genes, female wolves should act to optimize their reproductive output and males should do the best they can to help their mates succeed at this. Thus, the female should maximize feeding and caring for her pups, and the male should maximize helping her do this. The male should procure as much food as he can for both his mate and his pups. (Contrary to this, in the ungulate world, the male strategy is to impregnate as many females as possible and hope for the best; wolves sometimes employ this approach as well [Stahler et al. 2013], but not often.)

On Ellesmere, I had the privilege of observing the behavior of the wolf parents as they fulfilled their evolutionary obligations. It was often quite a dramatic scene, especially when I ran my various feeding experiments. I had purchased seal flesh from the Inuit and was licensed to shoot arctic hares. Between those two sources, I always had enough meat for my study, and even a bit to nourish myself.

I conjured up one of my tests while watching the adult wolves return to the den area and feed the pups. I had often wondered how much food it takes to motivate wolves to return from their hunt and feed the pups. Thus, on July 9–10, 1992, I decided to run a test on a wolf that had started out hunting. I would incrementally feed the wolf until it was sated enough to decide to return and feed the pups. And what better a wolf than Whitey, the mother of three hungry pups.

My test began when both parents were at the den at 2135 hrs on July 9, and Whitey seemed to be hungry when she licked up to her mate, but he failed to regurgitate to her. Both then slept for about 2 hours; then at 0042 hrs on July 10 they headed northward for a hunt.

When the wolf pair and I on my ATV got about 2 miles (3.2 kilometers) from the den, about 0115 hrs, I took the hind quarters weighing about 34 ounces (950 grams) of an adult female hare and threw it to the male. Whitey sprang right in and grabbed it out from under the male, took it about 30 meters away, and devoured it all except the feet. Then she grabbed the feet and some fur in her mouth, scratched the ground, and walked away. The male tried to seek what little remains were left on the ground, but Whitey snarfed them up. All the male could do was scratch the ground.

The pair started heading north again as if to continue on the hunt, so at 0130 hrs I tossed the hare head (110 ounces or 300 grams) to the male. Whitey immediately rushed him and grabbed the head, took it to where she had eaten the legs, and consumed it after squat urinating. She then finished off the leftover hind feet themselves and at 0141 hrs started back toward the den. The male just lay and watched. Thus, Whitey wanted to continue hunting when she had eaten 34 ounces (950 grams), but after consuming another 11 ounces (300 grams), or 1.25 kilograms total, that seemed enough to motivate her to return to the den. (I later weighed similar pieces of another adult female hare to obtain the estimated weights of those parts.)

One cannot generalize from this single, rough test, and the amount might vary with the distance from the den or several other factors. However, it does give some idea about the amount of food a hunting wolf might obtain before being motivated to return to feed the pups. And return to the pups is just what Whitey did.

My notes: "When 50′ [12 meters] from the pups, Whitey whined, and the pups rushed to meet her and Whitey regurgitated to them. Then she went 20 meters E with pups following and regurgitated a second time. Then she went to a lone pup on the rocks and regurgitated to it and back down and regurgitated a 4th time. These 4 regurgitations were all from the hare head and hind quarters. . . . When the male arrived [a few minutes later], Whitey licked up to him and he regurgitated to her and then to the pups. Whitey then went west ~50 m W of den and cached something." Four regurgitations of 1.25 kilograms of food averaged 11 ounces (313 grams) per regurgitation.

Not only did my experiment give me some new information about how much food might motivate a wolf's return from the hunt, but it also demonstrated how the male must passively allow the female to usurp food that could have been his had he fought for it. But the male never had a chance.

I saw other situations that were similar, even when the male caught the food himself. During July 15–16, 1996, I followed Whitey and her mate from the den as they left to hunt, so I could get some idea about what they accomplished before returning to the den. Whitey and the breeding male left the den area about 1814 hrs, and Whitey was back by 1855. This was not unusual because it seemed that Whitey often had to motivate the male to go hunting, and when he was well on his way, she returned to the pups (chapter 20).

This time the male visited an old muskox skeleton, spent about 40 minutes at the weather-station landfill, and then headed out to catch leverets. He caught leverets at 2100, 2315, and 0049 hrs and cached each a few minutes after he caught it. He caught a fourth at 0114 hrs and ate that one. At 0136 hrs, he caught another, which he then carried toward the den. I lost track of him at about 0200 hrs while he was still chasing leverets, but at 0359 hrs the wolf arrived at the den carrying two leverets.

I dug up and weighed three of the leverets, which averaged 47 ounces (1.32 kilograms), so presumably the one the male ate was about 1.32 kilograms. The wolf had also rummaged at the landfill so must have got an unknown amount of food there, plus he visited a putrid muskox carcass but didn't eat much there. Thus, this wolf was gone almost six hours, caught at least six leverets, and ate at least 1.32 kilograms before returning to the den.

As the male was returning, Whitey headed about 150 meters from the den to meet him and instantly snatched the hares. One dropped and she took the other about 6 meters and began eating it. The male picked up the other, but

Whitey charged over and seized it from him. The male then continued to the pups, which licked up to him, and he regurgitated to them at 0344 hrs. The pups and the male then ate the regurgitant, with the male eating for 2 minutes.

Whitey carried the second leveret (after consuming the first) to the pups at 0348 hrs, and one pup followed her, leaving the male and the regurgitant, while the other pup continued to feed until 0350 hrs. The male returned later to the regurgitant and ate for a few seconds. At 0355 hrs he regurgitated to one pup and consumed it with the pup. At 0408 hrs the male departed north and slept on the north slope while Whitey and the pups snoozed in front of the den.

It was quite clear which was the boss in that household, or at least which wolf was in charge of the family food. Still, even though the breeding female reigned supreme in that respect, I wondered about which wolf really called the shots when it came time to actually venture out and fetch the food. Until the pups were weaned, of course, their mother had to hang around the den longer and more often just to nurse them. But how did all that work after weaning?

As I wrote in an article on the subject (Mech 2000b, 260):

> Discerning leadership in activities preceding travel away from the den was complex. Usually, the activities included awakening of individuals, their awakening of packmates, considerable socializing, and eventually travel away from the den. Even after travel began, the trip could be aborted kilometers away when a pup or pups followed, and the breeding female eventually led or carried them back to the den.
>
> Generally, pack members awaited the awakening of their parents before becoming very active, although sometimes their activity would awaken the parents. Nevertheless, it was not until the parents were awake and active that much socializing went on. Often the breeding female awoke first and tried to awaken the male. Furthermore, the female sometimes seemed to urge the male to become active and go foraging. She would lead the male away only to have him lie down again, and the two would then begin howling. After that, the two would arise and go off again, but sometimes they would repeat this behavior a few times. Eventually the pair would leave the area, and after 5 to 30 minutes the female often returned alone, as Murie (1944) also observed, apparently having sufficiently motivated the male well enough to trust that he was actually continuing on. . . . Each summer, as the pups got older, the female tended to accompany the male and the rest of the pack for much longer periods. Of 29 times [during 11 summers from

> 1986 through 1998] that I was able to determine which Wolf led the Ellesmere pair or pack away from the den, the male was first 22 times and the female 7.

This difference was statistically significant.

Those observations made it seem like the breeding male was more of the leader because he led the pack on their travels away from the den more often, and generally tends to lead hunts and be more aggressive during them (Mech 2000b; MacNulty, Mech, and Smith 2007). This conclusion squares with findings with the Baffin Island Wopemado Pack in which the breeding male "assumed the dominant role" in all of the 67 interactions recorded between him and other packmates (Clark 1971, 100). On the other hand, the fact that the Ellesmere breeding female at times seemed to lure the male away and motivate him to hunt, plus her sometimes actually leading the pack away on the daily hunt, shows she also possessed strong leadership tendencies.

Where the breeding female unequivocally displayed her take-charge abilities was during a serious issue with a bull muskox that stood defensively at the den entrance with the single 5-week-old pup inside on July 11, 1994. The entrance was at the north base of the den's rocky ridge. which tailed off some 50 meters behind. The base of the ridge face, where the den entrance happened to be, formed an excellent defensive position for the muskox.

The muskox had wandered into a meadow just east of the den ridge when all four wolves (the breeding pair, and two 2-year-olds) noticed him and surrounded him. At some point, either he charged one of the wolves or they charged him, and there was a flurry with the bull running and bellowing and heading around the rocks in front of the den entrance and standing there (0220 hrs).

At about 0230 hrs, the wolves were frantic and harassing their intruder from all sides, including standing above him on the den ridge. With Whitey bark-howling, all four adults suddenly headed south around the east side of the den ridge and then back north on the west side. Until then, the wolves had trapped the creature right in front of the den entrance. But when the pack filtered around to the west side, all in unison with Whitey around 9 meters west of the den and still bark-howling, the east was clear to the muskox. From then on, the four wolves acted together, settling west of the den entrance for around 10 minutes, with Whitey closest. The breeding male took no particular lead in all of this. Whitey seemed to have done the trick.

Eventually, with all the wolves on one side, the muskox slowly moved east from in front of the den and sauntered back into the east meadow and began grazing (0315 hrs) around 25 meters northeast of the den but gradually heading northeast away from his harassers.

Thus, under various circumstances, it looks like either the breeding male or the breeding female can assume the initiative, which is pretty much what I concluded in 2000, calling pack leadership a "joint function" and also in a later study of three Yellowstone packs. The conclusion of that study was: "We found division of leadership to be about equal between dominant males and females, at least in winter" (Peterson et al. 2002, 1410).

With the breeding pair dominant to their offspring, the question about leadership in basic wolf packs, like those on Ellesmere and most other places that contain primarily a breeding pair and their offspring, is perhaps less complex than in areas like Yellowstone. There, several packs include two or more breeding pairs, the original breeders with pairs of their mature daughters and their mates. The Yellowstone folks are now pretty convinced that females hold "the highest leadership role" in packs there (Stahler et al. 2020, 48). See also McIntyre (2019, 2020, 2021, 2022).

Some of the disagreement and current lack of understanding about wolf-pack leadership possibly results from what the actual definition of leadership might be. Is leadership based strictly on which wolf leads the pack away from the den each day? While traveling when away from the den? While hunting? When returning to the den? While initiating other activities? While feeding offspring? Packard (2003) discussed these questions in detail while not necessarily answering any of them definitively.

One important area of wolf behavior in which males do seem more dominant is in fighting wolves from other packs. In Yellowstone, males, especially older males, were most effective in battles with other packs (Cassidy et al. 2017).

What about behavioral dominance? Which member of the adult pair is usually dominant? And does dominance imply leadership? I had already produced evidence that with my Ellesmere wolves, at least during summer, the breeding male tends to dominate all the other pack members (Mech 2000b). Most of my evidence was based on packs with offspring. For example, several years later, in 2009, the breeding male dominated one of the other males for at least 6.5 minutes while my companion videotaped it (Mech and Cluff 2010).

Such dominating behavior between the breeding male and an offspring, of course, does not in itself mean that the breeding male dominates all members of the pack, including his mate. However, I also had one year where the only wolves in my study area were the mated pair. What was the relationship like between members of this pair of wolves that had no offspring that needed feeding, a pair that had not yet bred or had lost their pups? Is the male or the female the leader of the pair, or is neither? How do the two adult wolves interact? No other wolf researcher had published any information about this, and my only observations about this subject came from 1998, when I was only in the study area for about two weeks.

Still, I did record some interesting observations during that time. I summarized them in my 1999 article on dominance and my 2000b piece about wolf leadership (cited earlier), but the following are my actual field notes about interactions between Explorer, the 6-year-old female, and her mate:

> July 2—At 0058 [hrs], Explorer licked up to the male in a standard submissive posture.
>
> July 7—She groveled low on the ground and licked up to him very submissively. He did not raise tail, just stood there. . . . if he got too close, she snapped and lunged at him from a very submissive posture.
>
> July 8—for 1 1/2 min. she did very elaborate submissive display, tail down and under, rump down, licking up, ears back, sometimes back of head to him, and once she turned around in crouched position with lowered rump towards him. This display was similar to that of pups and yearlings toward the adults but was unusual in its long duration. The male meanwhile just stood there with no obvious change in demeanor, not even his tail up.
>
> July 9—She approached the male at 0543 [hrs] and licked up to him and submitted to him as usual but only for 20 sec.

All these observations made it seem that Explorer was submissive to her mate and that thus he was dominant, similar to Clark's (1971) findings cited earlier. But this was just a sample of one wolf pair. What were the interactions between members of other wolf pairs? The only other person in the world who I knew might have similar observations was Rick McIntyre in Yellowstone. When I consulted Rick, I learned that in a similar pair of wolves without offspring, the interactions were such that it was not at all apparent if either member of the pair was dominant (McIntyre, personal communication by email, September 29, 2021).

Why does any of this matter? In the Big Scheme of Things, it really might not. Dominance is a human concept, so it might not even apply to members of wolf pairs. The Ellesmere wolves each had different personalities, so it could well be that whether a male or female wolf is dominant or tends to lead may vary with the particular pack, time, and place. So long as the members function well and survive and reproduce, that is what counts. Each pair member has its important role, with the female producing and tending the pups and the male securing food for the female and the pups while the female is nursing them (Mech, Wolf, and Packard 1999). In an Alaska study the breeding male provided more food to the pups than did their mother (Ballard et al. 1991).

As already discussed, evolutionarily the male's imperative is to feed his mate to maximize her ovulation, the number of his pups, and her nursing and feeding them. It is the only way the male can help get his genes reproduced and ready to perpetuate themselves. He continues to feed the female through early denning and until the pups are weaned.

The female is well aware of the male's imperative too. I found this out dramatically during the July 9, 1992, feeding test described earlier to see how much food a wolf would have to obtain before heading back to feed the pups. I also witnessed it the day before, when I tried to see how the breeding male and breeding female would act when away from the pups when I offered them a complete arctic hare. My notes tell that story:

> I threw a whole female arctic hare [carcass] to the male from 15′ and he grabbed it and started running with it. Instantly, Whitey [his mate] dashed in and grabbed it from the male who easily relinquished it. Whitey took the hare straight back to the den.
>
> I then took the second female arctic hare and similarly gave it to the male. He grabbed it the same way and carried it 75 meters, ate the head and then carried the hare straight toward the den. I zoomed around on the trails and got to the ridge E of den in time to watch the male arriving. He took the hare straight to Whitey who was at the base of the den rocks eating the first hare. The pups were in the rocks at front of den. Whitey instantly grabbed the 2nd hare, cowering [active submission] or deferring with tail down and even barked as she did.

She later cached the hare, showing that it really was not her hunger nor the pups' hunger that motivated her to grab the hare. It was her *right*, by golly!

A few days later, Whitey's mate, however, did once turn the tables. At 1832 hrs, when I could get both the male and Whitey together, I threw the male about a kilogram of seal meat. Whitey chased her mate some 25 meters but did not get it. The male devoured the meat quickly and returned. I then pitched the male a similar-sized chunk of seal, and again Whitey tried to grab it away, although less aggressively, and failed; the male ate it. On a third test a few minutes later, I gave the male about 17 ounces (0.5 kilograms) of seal, and he let Whitey get it. Whitey had eaten an estimated 50 ounces (1.5 kilograms) just before these tests, so that might have reduced her aggressiveness.

That reduced aggressiveness was only temporary, however. A day later, I ran a similar test. On July 12, at 1730 hrs, "I threw the male a seal bone with meat to test if Whitey could still usurp it; she charged over to him as he ran to the valley and tried to get it from him. He kept circling away from her, keeping his head turned away and she kept circling after him. They circled 11 times, and she got the bone and ate it."

To reiterate what I make of all this: feeding his mate and the pups during this period while the female is nursing the pups is the male's best evolutionary strategy to foster survival of his genes. The female's best strategy is to secure as much food as she can while she can, for the male will soon switch his approach from feeding his mate to feeding the pups. The female has done all the eating she can to help promote his genes once the pups are weaned. From then on, it is a better investment for the male to feed his pups directly rather than through the female.

Switching more to feeding the pups after they are weaned would be more efficient for the male to maximize his reproductive output. Thus, theoretically he should tend to forsake feeding the female until the next breeding cycle and focus on ensuring that the pups do well. Similarly, the yearlings and any other siblings of the pups best enhance their genes by feeding only the pups at this time. Those pups share some of their genes. Feeding their mother or their father would be a less efficient way to further their own genes. This is why the breeding male is the only member of the pack that does not get fed by any other pack member, as already mentioned. He has served his "purpose."

I had only two feeding test results during the postweaning period, which is roughly the end of July. At 0921 hrs on July 26, 1992, with both Whitey and the male within 3–4 meters of me and Whitey within 3–4 meters of the male, I lobbed a hare torso (no head, front legs, or back legs) weighing about 7 pounds (3 kilograms) to them. I had hypothesized that, contrary to the period when Whitey had been nursing, the male would not give it to Whitey or let her have it. That is what happened. The male snatched the hare right away, and Whitey ran to him and chased him about 30 meters but got only the guts which fell out. The male ate all the rest, returned to where the hare had landed, and RL urinated.

A few years later, I tried another test, this one on July 31, 1996. Here are my notes:

> [Whitey] . . . resumed the submissive posture and ran to him [her mate] and licked up at 0907 [hrs] but he did not regurgitate. After several minutes I let them start heading toward the den and decided to test the male to see if Whitey could still take food away from him. I threw him the side of the chest of a hare, but Whitey beat him to it, although clearly he deferred to her aggressive approach to the piece. While she was eating that in one direction, I threw the male the other side in the other direction, and he ate that. They both headed slowly to the S end of the den rocks, and there at 0920 [hrs] Whitey again licked up to the male but he did not regurgitate.

Two 1990 observations also support my hypothesis. On August 1, I watched as the returning breeding male regurgitated to the pack's single pup and while both Mom and Whitey begged from the male, the male merely snarled at them. Four days later, the male regurgitated to the pup, and Mom tried to eat some of it, but the male snarling at her prevented that.

The breeding male even carries his imperative so far as to feed his post-weaning offspring when they are yearlings and no pups are available, as I observed in 1993: "Four times when the pack was observed hunting arctic hares, the alpha male ambushed the leverets being chased by yearlings and killed the hares. Three of these times he either immediately dropped the leveret in front of the yearling and the yearling consumed it, or he quickly relinquished it when the yearling approached; once . . . he gave it to a yearling after 45 minutes of apparently waiting to give it to the alpha female [who failed to appear]" (Mech 1995a, 474).

These observations all accord with inclusive-fitness theory (Hamilton 1964), so the breeding male's behavior is not surprising. What would be surprising would be if the breeding male fed some unrelated individual, and I never witnessed that. In fact, to my knowledge, there is only a single record of that, although it is an interesting one. In Yellowstone National Park, a breeding male fed a nine-year-old crippled female which, so far as is known, was not related to him. "I caught sight of her just after 685M, the breeding male, dropped an elk leg onto her forepaws. 685M had pulled the leg from the carcass, climbed the hill to where the old female lay, and brought her the meat," wrote Ilona Popper (2017, 24).

That latter bit of altruism is harder to explain from an evolutionary standpoint and still stands as an isolated record. Generally, the breeding male's feeding of his mate and offspring are fine examples of kin-selection (Hamilton 1964).

When I published my 1999 article, "Alpha Status, Division of Labor in Wolf Packs," I concluded that "the typical wolf pack is a family, with the adult parents guiding the activities of the group in a division-of-labor system in which the female predominates primarily in such activities as pup care and defense and the male primarily during foraging and food-provisioning and the travels associated with them" (Mech 1999, 1196). Now, more than 20 years later, I do not feel I can improve on that.

: : **19** : :

Life at Wolf Headquarters

For most of the year, the Ellesmere wolves, like those throughout the world, live a nomadic life within their territory. I was long keenly aware of this, having radio-tracked wolves in other areas since 1968. Only when raising pups do wolves headquarter around a den, during late spring and early summer in most areas. Still, during that short proportion of the year-round wolf life, that is when so many of the most interesting and intense interactions among pack members take place. And in most wolf-study areas, that is when it is hardest for anyone to see or study them—except now with remote cameras, or in real time on Ellesmere.

True, Yellowstone and Denali National Parks and some Canadian national parks sometimes offer such an opportunity, although certainly not 24 hours a day. And not with such cooperative subjects as the Eureka wolves. That is why Ellesmere was such a dream for me that I seasonally forsook my other projects and focused on it for parts of 24—almost 25—summers (fig. 19.1).

Wolf life around the Ellesmere dens, however, was not a constant circus. Far from it. Nor was being there watching the wolves around the den day after day and night after night as romantic and stimulating as folks might imagine. When giving talks about the project, I was fond of dispelling some of these notions by describing a typical wolf-watching session this way: "You sit on an ATV all decked out in heavy arctic gear, blasted constantly by frigid wind, watching wolves sleeping for 12–15 hours a day just hoping something interesting will happen." Then I add that I got lots of books read, often dozing off myself.

Figure 19.1. Being able to observe wolves 24 hours a day was such a dream for me that I seasonally forsook my other projects and focused on it for parts of 24—almost 25—summers.

So maybe I was exaggerating a bit, just for effect. But, yes; wolves around the den do sleep a lot. And why not? When they are awake, their long travels, often over difficult terrain, expend plenty of energy. For example on July 22, 1994, I wrote: "the wolves all went to the N. meadow where the alpha male just flopped on his side and slept as though very tired. He had been gone about 66 hr."

One could say that about wolves almost anywhere, but it is especially true of Ellesmere wolves because their main prey, muskoxen, live in herds that are literally few and far between, scattered across the landscape. Thus, the wolves' hunting jaunts cover large areas in a hurry. And the long-legged canids might be gone for long periods.

In July 1988, for example, the breeding male left the den area for over two days, during which he might have covered hundreds of kilometers. When he returned, he regurgitated large piles of meat, so his hunt had been successful. Thus, when these peripatetic creatures can rest, they take advantage of the opportunity and "crash." I described the seven 1986 adults on July 16, the day after a long and successful hunt: "they all looked completely beat—mouths open—which we had never yet seen, and haggard, worn out and exhausted." I have watched wolves sleep for as long as 16 hours at a time (for example, Whitey, July 11, 1988).

Such long sleeps do not seem unusual for an animal that may travel 12–18 miles (20–30 kilometers) daily, for even dogs, whose lives are subsidized by humans, generally sleep 10–12 hours per day. Because dogs and wolves are so closely related, it is reasonable to think that their sleep characteristics are pretty much the same. Only a single study has compared them and found that they are fairly similar, although young dogs "appear to have spent less time in REM and more time in NREM [non-REM] compared to young wolves" (Reicher et al. 2022).

Of course, not all wolves sleep in long bouts, for hunting habits vary among different wolf populations. Way farther south, for example, in forested areas where there are more prey and a wider variety of it, continuous travel and long bouts of sleep would not be needed. Wolves can find some prey there much more often even if they still have trouble killing it. Thus, their travel routine is much different from that on Ellesmere.

For example, in Poland's Białowieza Forest, wolves were active for an average of only 45 minutes at a time. But they have a much denser prey population there than do the Ellesmere wolves: red deer, wild boar, and roe deer with mean densities of about 8–18 red deer/mile^2 (3–7/kilometer^2), 3–15 wild boar/mile^2 (1–6/kilometer^2), and 3–13 roe deer/mile^2 (1–5/kilometer^2) (Theuerkauf et al. 2003). Being active in only short bursts mean wolves only sleep in short bursts. "Most active and inactive bouts lasted only 15 min (46% and 37% of bouts, respectively) or 30 min (18% and 16%, respectively), whereas only 19% of active and 30% of inactive bouts were longer than 1 h. The longest activity bout lasted 7 h and the longest bout of inactivity 9.5 h" (Theuerkauf et al. 2003, 247).

Conversely, muskox herds or individuals occur in densities better measured in number per 1,000 kilometers^2 on Ellesmere. Then, if hunting-success rates by wolves are generally as low as I estimated them to be during July (chapter 15), just to find one herd in which they can kill a calf, Ellesmere wolves must locate about 5 herds on average. To find an adult they can kill, they must run into about 24 herds and/or singles. Doing so amounts to a great deal of travel, which uses considerable energy and explains why they sleep long and hard.

It has been quite apparent over the years of observing these wolves around their den that the default behavior is for both adult parents to frequent the den area and interact with the pups just about every day. However, there have been times when that has not been possible. Perhaps one day all the muskoxen the wolves encountered had already been tested recently so the wolves had to continue on beyond them to try to find others. Or maybe the herds that the wolves expected to be in certain areas had shifted to a new area. In any case, sometimes the adults don't make it back to the denning area each day.

Generally, breeding females spend the most time around dens, while their mates are out hunting. In Idaho, where wolf homesite attendance by various adult pack members has been well investigated, the amount of time each wolf spent away varied considerably. For each additional nonbreeding adult in a pack, each wolf spent 7.5% more time away, and the nonbreeding males were gone longer than the nonbreeding females (Ruprecht et al. 2023). The latter finding could be construed as behavior precursive to each gender's ultimate behavior as breeders.

On Ellesmere, it's likely that the time both parent wolves were gone the longest was in 1996, when there were no other adults in the pack and only two pups. The adults' long absence started on July 17 at about 1902 hrs when the pair headed southeast from the den, a common direction that led in about 5 miles (8 kilometers) to the Eureka Weather Station, airstrip, and military base along the north shore of Slidre Fiord. The semblance of an old roadway paralleled the shore from there eastward for some 7.5 miles (12 kilometers), a favorite wolf route. Beyond that for many kilometers lay sandy desert and only scattered vegetation. But if the wolves veered north at the end of the roadway, that took them around to the east side of the 6 mile (10 kilometer)–long Black Top Ridge and much more vegetation.

That is apparently what they did, for my partner and I on ATVs followed the pair from a distance, and at 2257 hrs the wolves were 15 miles (24 kilometers)-travel distance from the den and seemed to be heading north along the east base of Black Top. I wondered then about this journey and speculated in my field notes as follows:

> During this trip, the wolves were traveling very intently, not stopping to hunt leverets, even though they passed several adult hares, and one jumped a leveret along their route. This makes one wonder what this trip was for. They could have hunted leverets in many areas much closer to the den, and even musk oxen much closer. Possibilities: (1) saving local leverets till later (2) going to new areas to hunt m.o. not recently

> sensitized (alert) to their presence (3) marking their territory, especially the female who had been fairly restricted by pup care.
>
> It may be significant that the pair took this trip when they had a fair supply of food. The night before, the male was known to have cached three leverets, and the female two, and that is what we know of. The male could have cached more last night without our knowing it. Last night the female had eaten from 2 other caches. They also had the old muskox carcass with much meat left, although I'm not sure how much of that they could get at.

I never did satisfy my curiosity about why the wolves were making such a long journey that took them so far from the den for so long. After watching the pair disappear that night, my partner and I headed back to the den and our camp. We continued to watch for the pair to return throughout the next day and then the next.

Finally at 1242 hrs on July 19, I found that Whitey was just back and nursing the pups. She also regurgitated to them, and I thought I saw pieces of both muskox and arctic hare. However, Whitey also appeared to be hungry herself. She ate some of what she had thrown up and then dug up a cached leveret and ate it. She saved a piece to take to the pups, which made me wonder, Does she carry a piece to the pups, so they don't make her regurgitate the rest?

Thus, I decided to test how hungry Whitey was. I tossed her a hind foot and lower hind leg of an arctic hare, with most meat removed, and she quickly crunched up even the hind foot and ate it. The hind feet of hares hold no meat, just fur and bone, and wolves generally don't eat them unless really hungry.

The male too was hungry upon his return at 1438 hrs the same day. Whitey and the pups rushed him as usual, but they all went out of view, so I couldn't see whether he regurgitated anything. However, they all quickly returned in view as though he hadn't. He also had dug up a seal flipper he had cached in the den area a few days before and ate that. As a final test of how hungry he was, he eagerly crunched up a hare leg bone (without most meat) that I tested him with.

Thus, even though the breeding pair had spent about 41 hours away and presumably hunting most of that time, they seemed to have little to show for it. Considering that they had traveled 15 miles (24 kilometers) in about 4 hours, if they only spent half of their time away traveling and half resting or sleeping, they still could have covered over 60 miles (97 kilometers) searching for prey. And come back hungry.

Meanwhile during most of the wolves' time away, I would have been observing at the den with only the pups there, and they sleep a lot too. So I got some good reading in.

On other occasions, while I was watching there at the den, the wolves would burst into interesting activity of unknown significance, as on July 23, 1994, just after midnight:

> 0055 hr[s] Whitey [the breeding female] and the pup were playing with the seal skin and Explorer [2-year-old female] went running to them. Whitey suddenly pinned Explorer severely and Explorer rolled over in extreme submission for 1–2 min. with Whitey very assertive. Explorer licked up to Whitey and the pup licked up to Explorer all at the same time as they flailed around on the ground with Whitey crouching over Explorer and Explorer on her back and side.
>
> A few minutes later, Whitey abandoned the seal skin. Explorer then mouthed it for 1–2 min. and then dropped it. "What was that all about?" I wondered.

Such is the life of a wolf. And a wolf biologist.

Much of the activity around the den involved the pups developing little by little and expanding their horizons. Besides the usual daily investigating of each other through play and wrestling, the pups crept farther and farther from the actual den hole itself, exploring their surroundings, every little nook and cranny of the rocky ridge that housed the hole. The adults themselves also had to keep their eyes on the pups as the little furry critters began developing into miniature adults and ventured away from the protection of the den ridge and out onto the tundra.

At first the hummocks and crevices of the uneven ground thwarted the pups' attempts to negotiate them, but gradually the growing canids conquered them and seemed to delight in tumbling around and over them. Eventually, the pups realized they could quickly stray away into the wider world, but then the adults would have to nudge them back to safety. During my very first year on Ellesmere, I was impressed by how the adults did that (July 20): "The adult put her nose against the rear end of the pup and nudged it [along] for about 50 feet up the hill to the den."

In some cases, the adults, usually the breeding female, had to pick up a pup and carry it back to the den, sometimes for many meters. As the pups grew and developed, especially in late July and early August, they tended to follow the adults when leaving for the daily hunt, and the adults might have to spend much time herding them back to the den or rendezvous site.

Another handy way adults used to direct the pups to where they wanted them was to pick up a bone or other toy and lead the pups there. The pups were always eager to follow and eventually get to play with the toy. The adult could then settle down to rest or sleep in preparation for another daily trek away to hunt.

The next foray the breeding pair makes might just end up completely opposite of the long, not-too-fruitful trek described earlier. A good example of the resulting activity around the den after the pack makes a large kill comes from a few days' worth of my 1994 field notes when the pack consisted of Whitey, Explorer, the breeding male (then called "Alpha male"), and 2-year-old male Grayback II; and a single pup.

It was clear that the pack had recently killed a muskox not too far from the den. It turned out to be about 3 miles (5 kilometers) away. Companion Craig Johnson and I found the kill on July 12. It was on the riverbed along the base of the west bank of Station Creek. Grayback II was lying 2 meters above it on the bank while the breeding male and Explorer were eating. At least twice, Explorer cached pieces.

At 1813 hrs Grayback II left the bank, I thought maybe because of our presence some 8 or so meters away, crossed the river but did a RL urination as he did, and lay along the opposite riverbank around 100 meters from us.

The breeding male took around 1/4 of the liver several meters up on to bank at about 1820 hrs and ate it, then returned to the kill.

About 1825 hrs, Explorer and the male stopped eating and headed over the hill to the southwest and disappeared, presumably going back to the den. Grayback II then proceeded to the kill, and after we backed off a bit, he began feeding at 1832 hrs. I thought it significant that he could not feed until Explorer [female] and the breeding male were done. Was this a male-versus-male competition, I wondered? Would Whitey have let him eat with her? Would she let Explorer eat with her?

Grayback II tore out something around 4 pounds (2 kilograms) from the kill and at 1834 hrs carried it to the east bank of the river and over the hill east of the river, presumably to eat or cache it. He reappeared at 1852 hrs about 400 meters south. and returned to the kill at 1855 hrs and ate until 1916 hrs, when he moved west over the hill.

I examined the carcass and kill site. There appeared to have been a scuffle on the top of the 2-meter bank but no blood. Then at the base of the bank below the scuffle area was a pool of blood, including clots, and a place where the muskox had lain in mud against the bank. A small area of mud only 2 meters from where the wounded animal had lain was packed with wolf tracks as though perhaps the wolf or wolves had bitten the muskox while it was lying there. Had the drop over the bank injured the creature enough so it did not fight hard? Or was it in such poor condition that it could not fight?

The muskox apparently had then moved around 3 meters and fallen where the carcass lay on its right side. It was about half eaten (all of the left side, including its left hind leg and pelvic area, but the lungs remained). One fist-sized testicle was lying naked next to the remains, and I confiscated it. My notes divulged what I later did with it: "Sliced up muskox testicle, fried and ate it. OK but not especially tasty."

Based on my records of the times wolves were at the den, I figured that this muskox must have been killed July 11 in the evening by the breeding male and possibly Grayback II.

The day after Craig and I examined the kill, the wolves' activity around the den reflected the sudden abundance of food:

> At 1433 hrs, Explorer arrived from the E . . . and headed to the den, got the pup and led it 50′ S but no regurgitation. The pup did a STU. Explorer led pup another 15′ and regurgitated to it. The pup ate it, and Explorer also re-ate it. (Explorer, whose muzzle was quite bloody yesterday, was almost clean today.)
>
> Whitey returned from the NE at 1530 hrs and Explorer licked up to her, but Whitey did not seem to regurgitate. Both Whitey and Explorer went to the pup. I saw no regurgitating but they were out of sight much of the time.
>
> 1630 hrs—Grayback II arrived from the NE and regurgitated to Explorer and the pup. Whitey remained lying and did not beg or look up. Grayback II went to Whitey, who was lying with a meat chunk, and Whitey snapped at Grayback II. Then Whitey took chunk 10′ and cached it lightly. Explorer quickly took the chunk, ate some of it, and took it 100′ SE and cached it and returned to Whitey, the pup and Grayback II and all slept.
>
> 1704 hrs—Whitey went to Explorer's cache and spent around 3 minutes looking for it but could not find, came back to Grayback II and stood over him 1 min. and lay near him.

1721 hrs Whitey took a meat chunk (around 5 lb) S and out of sight.

1749 hrs saw alpha male regurgitate into a cache in a muddy sedge meadow near 20′ wide pond around 1 km S of the den. I fed him. He then went N toward the den. Craig saw him arrive there at 1800 hrs. Explorer and Grayback II licked up to him, but he went to the pup and regurgitated. Explorer got some. Grayback II had already walked away.

Whitey returned at 1800 hrs and licked up to the alpha male. The pup tried to suckle but Whitey walked away. Not sure whether it nursed or not. Then all slept except Explorer. . . .

She then headed W and at 1832 hrs returned with a chunk of bone meat (around 3 lb.) and dropped it to Whitey and pup but neither showed interest. She then slept.

In this example, the key was the adult muskox the pack killed so close to the den, one of three such adult-muskox predations I recorded during my 24 summers. This close kill allowed the wolves to ferry back and forth between pups at the den and the massive food source. Thus, the unusual amount of activity and the lack of long naps for the adults. And little book reading for me.

:: 20 ::

Just for the Howl of It

> July 31, 1991–2304 hr–Whitey started howling, and BM and Mom joined in, and all ran excitedly to each other, and mouthed each other solicitously but no pinning, and all looked S and sat within a few feet of each other and howled to the S [4.6 kilometers across a fiord]. It looked very much like they were responding to an outside threat to the S. . . . Whitey and Mom continued howling to 2316 hrs.
>
> 2330 hrs—Whitey howled and looked SW. She kept howling, and at 2333 BM joined, and all howled till 2343 hrs.

So I wrote about trying to figure out why these wolves were howling so persistently.

I could hear no distant howls, but clearly Whitey and her companions could. These three wolves were the only adults in the pack that year, so the only howls they could have heard were from a neighboring pack. That incident was one of only a few times I heard the wolves engaging in what was almost certainly a territorial advertisement (Harrington and Mech 1979).

Territorial advertisement is one of the four "supposed" main functions of wolf howling, with the other three being assembling (Murie 1944; Mech 1966), locating mates (Harrington and Asa 2003), and social bonding (Crisler 1958). Bark-howling is an additional type of howl that seems to function as an alarm or threat (Schenkel 1947; Joslin 1966). In addition, it has been pretty well established that wolves can identify each other via their howls (Palacios, Font, and Marquez 2007).

Logistically, it has been too hard to experimentally test wolf-howling functions, so most such information is based on inferences from anecdotal observations such as I was making. A study of more than 10,000 spontaneous wolf howls in Yellowstone over a 10-year period focused on the timing of the howls and other descriptive details but added little new information related to the actual functions of howling (McIntyre et al. 2017). It did confirm earlier findings that throughout most of the year, much of the howling emanates between packs.

During summer, however, most howls derive from members of the same pack, and that was my experience on Ellesmere as well. Because my Ellesmere studies all took place in summer, most of my howling observations consisted of a few lines in my notes such as "July 30, 1996; 1447 hrs- The pups headed N up the N valley and Whitey and the male howled and the pups returned to Whitey. All 4 headed to the NW slope. Whitey and the male lay down."

Nevertheless, I did experiment a bit with my wolves' howling, and I often watched more complex interactions among these wolves that gave me new insights into how howling functioned among members of these packs.

Several times, for example, I saw either of the individual breeding wolves seemingly use howling to motivate the other to head out to hunt. That may have also been what Murie (1944) observed when he described a wolf pack all getting together, wagging tails and chorus howling just before taking off for a hunt.

On Ellesmere, such a function became quite apparent on July 10, 1996, for example, when the breeding pair were the only adults. Just after midnight, the male Left Shoulder did a "silent" howl two times. He had been losing his voice as he aged, then being at least 11 years old, based on my annual identification of him (Mech 1997b). Although both my companion and I saw the animal lift his head and appear to howl, neither of us heard him even though only 30 meters or so away. The wolf then headed out of sight. At 0020 hrs, Whitey started howling, and the male joined in. This time the howls were real. The male headed north away from the den and then east along a valley and when about 300 meters away stopped and howled at 0027 hrs. Whitey replied, and they both howled.

A few minutes later, the male lay down and both he and Whitey continued to howl. Whitey was still lying with the pups near the den and watching her mate below. Was he calling her to join him? At 0034 hrs, she stopped howling, but the male continued while remaining prone. Whitey then stood, and the male stood. Both continued howling until 0039 when the male lay back down after walking only a few yards.

Two minutes later, the male scent-rolled. While Whitey moved a few yards to the den entrance, the male lay back down. A few minutes later, Whitey howled again, once looking at the male, and the male howled five times.

At 0047 hrs, the male started east again, and Whitey howled. The male stopped, replied until 0050 hrs, and then continued east. Whitey stopped howling, then arose (0052 hrs), left her two pups about 5 yards from the nearest den entrance, and ran down to the valley and east following the male when he was about 0.25 miles (0.4 kilometers) away. She scent-marked where he rolled and in other places where he had marked, howled at 0053 hrs, and caught up to her mate a minute or two later.

Both wolves then continued east with the male generally leading. I watched with my 10x binoculars as the pair continued due east for several miles. Sometimes they were one behind the other and other times side by side but 100–125 yards apart.

At 0106 they stopped and howled when about 2 miles (3.2 kilometers) east of the den, and I was amazed that I could hear them howl despite my ears being covered with a wool hat and my heavy parka hood and a stiff wind from the west. Why the creatures howled when far from the den and clearly in agreement to head out on the hunt, I still wonder. Were they disagreeing on the direction to go? I lost them from view at about 0140 hrs so I never found out, but that remains a reasonable hypothesis. Whitey did not reappear until 0441 hrs, when she delivered a leveret to the pups and began nursing them. I'm not sure when the male returned.

While writing my field notes about another apparent "argument," I even added my instant interpretation of what seemed to be going on between the two wolves. The occasion came 6 days later. The breeding male howled at 2122 hrs. Whitey joined in, and both continued for about 7 minutes. Whitey headed east down a valley about 200 yards from the den at about 2131 hrs while the male set out to the den.

Whitey then turned around and traveled back toward the male and howled once. They both traveled south and out of sight at 2133 hrs. Five minutes later I found them 150 meters southeast of the den. The male lay down on a hillside at 2140 while Whitey, a few meters ahead and below in a valley, disappeared.

At 2143 hrs I saw the male howl but couldn't tell which wolf had started the howling because Whitey was out of view. Both howled till 2149 hrs, and then the male disappeared. However, Whitey suddenly showed up near the den about 2153 hrs, having returned out of my view. A few minutes later the male also reappeared there and lay down. I wrote: "It appeared to me that Whitey was trying to get the male to go but couldn't get him to do it."

However, at 2204 hrs, the male got up and headed east, and Whitey ran after him, leaving the pups. The two adults then journeyed east down the valley, and when 300 meters away, Whitey led the male (2208 hrs). A minute later, the male turned northeastward while Whitey continued down the valley. There she veered north, and at 2213 hrs, she and the male howled while 100 yards apart parallel with each other. They continued howling for almost 10 minutes. Then Whitey stopped and struck out again, and the male howled once more and came over and joined her. They continued on down the valley, Whitey leading.

A few minutes later, the two resumed howling and continued for more than 5 minutes. Then they headed in parallel across some flats 115 yards (105 meters) apart. Whitey kept following the bank of a stream toward the southeast, whereas the male kept going east-northeast. They were soon 320 yards apart (2240). The male howled, and possibly Whitey too. (I could only see the male.) The male continued howling until 2242 and then veered south toward Whitey, and they both disappeared from my view.

"It seemed clear that Whitey had one way or destination in mind, and the male another," I wrote. "Twice, however, the male gave in and headed to Whitey. . . . At 2328 hrs, I spotted them 5 miles [8 kilometers] away . . . about 100 m apart."

My notion that the howling contest I witnessed above might signify a disagreement between the breeding pair is supported by a simpler observation that involved the pair heading out to a known kill. That happened on July 23, 1992. At 0446 hrs, the breeding male started howling, and Whitey and the pups joined and howled 32 times. There was no back-and-forth howling, just a common chorus. Then both adults together beelined it northeastward to the remains of an adult bull muskox they had killed several days before. No argument about which way to go then.

In another case, Whitey, then a yearling, on July 17, 1988, howled 21 times at 0140 hrs, but not one of the other three adults present that year (a yearling brother and their mother and father) replied at all. Thus, Whitey just headed off on her own and did not return until 1105 hrs.

Pondering all these and my many other observations of Ellesmere wolves howling, I concluded that a very basic function of howling is as a simple attention-getting device. "Pay attention to me; hear me out!" As such, it can be used for any number of purposes. A pup howls, and an adult comes over to it. A yearling or adult wants a companion to go hunting, so it howls. If no other pack member is interested or motivated, it goes alone, but if there is interest, that animal howls too.

The prehunt howling ceremony that Murie (1944) first publicized is the combination of all the adult pack members assenting to a group hunt. If members of the breeding pair can't agree on which way to go, each howls until finally one member gets its way. If the breeding female is hungry and wants the male to fetch some food, she howls, and if her mate agrees, he howls too. Or vice versa.

Other information that I learned about howling came from a few tests I ran and from miscellaneous observations. The tests I conducted usually involved posting someone with a two-way radio near the den when the adults were present. Then I drove several kilometers away, howled, and radioed that information to the posted person to see if any of the wolves heard the howls. If not, then I moved closer a set distance and repeated the process until the wolves heard the howl. After various tests under different conditions, the results showed that with relatively calm-air conditions, the farthest wolves could hear me howl and respond was 4.75 miles (7.6 kilometers) away.

As for how far humans could hear wolves is concerned, I got an inkling of that when, on July 13, 2009, at 0035 hrs, my companion, Dean Cluff, heard a wolf howling from across Slidre Fiord when the air was pretty calm. All I had to do was measure the distance across to learn the minimum distance in that instance, and it turned out to be 3.5 miles (5.6 kilometers) away. (My hearing by then was too poor to hear the wolf, but both of us with binoculars saw the white animal along the opposite shore.) We were with the Eureka wolves howling on one side of the fiord while the wolf across the way was howling, so presumably the wolves were hearing each other at that distance too.

I also got some idea how well wolves could pinpoint my location from my howling. On July 5, 1992, I posted my partner near the den with the adult wolves present and sped away to a point out of sight of the den 1.7 miles (2.7 kilometers) away and howled three howls at 1318 hrs. The breeding male responded with a howl and started sauntering "leisurely" toward me, followed by Whitey a minute later, according to the person at the den.

At 1331 hrs both wolves were about a mile (1.6 kilometers) toward me on a hilltop, and Whitey howled for about 5 minutes; then she headed back toward the den, arriving at 1350 hrs. Her mate, however, continued on, and at 1336 hrs cut behind a ridge east of me and reappeared about 30 meters downwind of me. The wolf then came to within about 6 meters of me and did a standing urination. It seemed like the animal had pretty well figured out where my howl had come from and decided to check it out even though I had not howled again or moved after my first bursts.

This instance came to mind when my companion Nancy Gibson and I were trying a distance test 10 days later.

July 15, 1992, happened to be one of the few days with calm air, and a BBC filmmaker, Richard, was with me for a few days then. I asked Richard to remain at the den with Whitey and the breeding male while my companion and I headed east to an area where we could communicate with him via two-way radios.

The idea was for me to radio Richard to watch the wolves while I howled from varying distances. Richard could then see whether the wolves reacted as though hearing the howls. When some 4.5 miles (7.2 kilometers) east of the den on a ridge from which I could see the den, I howled, and Richard noted that Whitey had perked up her ears, looked east, and wiggled her head sideways as though straining to hear. This was similar to her response when I had howled from 4.0 miles (6.4 kilometers) away.

Suddenly, three wolves came charging briskly from the east behind us, looking very intent. They meant business! They paid no attention to me and Nancy on ATVs standing out oddly on the barren tundra. No; they were searching for the wolf (i.e., me) doing the howling, and appeared almost bewildered that no wolf was in sight. Still, they frantically kept looking. The most serious and intent wolf was slim, like a female, and the other two appeared more thick-chested.

They were unafraid of me and Nancy but really wanted to get on with finding the wolf that had been howling. It was only through my driving around them and with them and tossing them bits of food that temporarily distracted them from their goal. Twice I got off my ATV and started to walk toward them. They quickly headed away but maintained their demeanor for about 45 minutes. When they were about 200–250 yards away, I decided to test them by howling. They did stop and look and lay down but did not approach. I howled two or three times more. After several minutes, the wolves went on, their search for the howling wolf seemingly defused.

Once again, I did not know which wolves these were, but because they were unafraid of me on my ATV or on foot and one or two responded to me like wolves I had worked with, I figured they were previous members that had dispersed from the large pack in 1989, which besides Whitey, Mom, and the breeding male, also included five other adults plus four pups.

I picked up a couple of other miscellaneous tidbits of new information related to wolves howling or hearing during my intimate interactions with the Ellesmere wolves, odd facts of little consequence at the time but interesting still and maybe important someday. For example, as already mentioned, Left Shoulder started losing his voice as he aged. On July 10, 1996, his two initial howls were actually silent. Again, the next day, I wrote the following about his voice: "Whitey slept 1504–1538 hrs. Then she howled 8 times and the male replied and both howled till 1552 hrs. The male's howl is so growly it sounds like a lion [roaring]."

This wolf was 11–13 years old, which is toward the limits of wolves' lives in the wild (Mech 1988b), and in fact I never saw Left Shoulder again after summer 1996. I did feature the old wolf in an article describing him chasing an arctic-hare leveret for 6–7 minutes, catching it, and then resting for at least 22 minutes (Mech 1997a) but did not mention that the old guy was also losing his voice.

The other interesting piece of howling-related info I learned was about Whitey's hearing ability, about which I wrote on July 7, 1996: "It [the air] was calm, so I could test Whitey's hearing. She could hear the slightest rustle of a white plastic bag at 125′. Each time I pressed the bag slightly, she turned and looked at me."

Admittedly this was not an earth-shattering finding. It did, however, exemplify the myriad little "factlets" that I assimilated as I bathed in the unique proximity of the Ellesmere wolves. After a career full of live-trapping their brethren in other populations, raising them in captivity, observing them from the air, gathering tens of thousands of their locations, and discovering information filling hundreds of articles and several books, I had finally come to know them in such an intimate way by embedding myself in their society.

:: 21 ::

Territoriality and Scent-Marking

Because during summer wolf scent-marking is at its nadir, I did not learn as much new information about this behavior as would have been possible during other seasons, other than the above about marking food caches. That behavior might even be considered incidental if the main function of marking is social. That is, it might just be that of the many different types of prominences where wolves normally mark throughout the year, food caches are just one of them. For example, wolves in Poland marked heavily around dens as well as around the edges of their territories (Zub et al. 2003).

Scent-marking increases throughout the fall and peaks in winter (Peters and Mech 1975) coincident with testosterone secretion and the breeding season (Asa et al. 1990), so one of its functions appears to be related to reproduction (Peters and Mech 1975; Rothman and Mech 1979; Asa et al. 1990; Ryon and Brown 1990). Marking behavior is thought to be dependent on testosterone plus either the presence of a mature, opposite-sex individual (Rothman and Mech 1979; Mech and McIntyre 2022) or dominance of a same-sex individual (Asa et al. 1990). Thus, other important functions seem to be advertising social status and, when along with marking by a member of the opposite sex, mate-possession (Rothman and Mech 1979; Asa et al. 1990).

Wolves, like many other mammals, possess a special organ just above the roof of their mouths, the vomeronasal organ (VNO), that conveys information from odors to the wolf's hypothalamus via the brain's accessory olfactory bulb (Harrington and Asa 2003). In other words, not only does information from urine, feces, and anal glands pass through the usual smell route, but also via this special passage that leads directly to a cranial center of reproductive and other behavioral activity. Thus, wolves sniffing scent-marks may be detecting all kinds of information other than that we have been able to discern so far. The anatomy of the entire wolf VNO system was recently elucidated in great detail (Ortiz-Leal et al. 2024), so hopefully we will soon have more knowledge about other kinds of information wolf scent-marks carry. Territorial advertising is a fourth function attributed to scent-marking (Kleiman 1966). Scent-marks are thought to be no-trespassing signs to inform neighboring packs where the edges of the markers' territory are (Peters and Mech 1975; Paquet 1991; Allen, Bekoff, and Crabtree 1999 [coyotes]; Zub et al. 2003). Recent information tends to support that function in that the RL urination marking rate tends to increase with the number of packs in a population, at least in winter (Thiel and DeWitt 2022).

Of those four putative functions, the two that would seem to be important even during midsummer would be social status and territoriality. Reproduction had already occurred well before and was underway then in the form of pup care and provisioning. The breeding pair were "well married" by then, although just which was in charge was still being determined (see above).

Nevertheless, I was able to add some information to what was known about all four functions of scent-marking. For example, I documented that, even during summer, free-ranging wolf pairs double mark (Rothman and Mech 1979)—a raised-leg urination by a male associated with a flexed-leg urination by the female (Mech 2006). And such marking was initiated about equally by female or male. I also found that breeding males marked significantly more often, as occurs with captive wolves (Mertl-Millhollen, Goodmann, and Klinghammer 1986; Asa et al. 1985, 1990; Nunez and de Miguel 2004; but see Ryon and Brown 1990).

My most significant finding involved the graduation by a maturing female, Whitey (an apparent yearling in 1988), to breeding status replacing her mother in 1990. I (Mech 2006, 469) wrote: "This pre-breeder began dominating the breeding female, her apparent mother, about the time the pre-breeder began FL urinating. . . . The pre-breeder bred the next year, apparently with her brother, and remained bonded to him for the next 6 years, producing pups during four of them while the mother remained with the pack for two more years without being seen FL urinating and then disappeared." This behavior was similar to that of a young male offspring starting to RL urinate and challenging its same-sex parent for dominance that I had observed in a captive colony (Asa et al. 1990).

A new finding that I did not even realize at the time involved multiple RL urinating in the same pack. That had never been reported in the scientific literature, but on July 31, 1987, as I was following my pack on a hunt, I noted the following: "In one place they did much scent marking on some rock formations at a prominent point. Two males RLU'd at once and 1 female SQU'd or defecated and it appeared that all wolves marked, some 2–3 times although I can't say for sure."

As I was writing this in 2022, I realized the significance of this observation, for only a few months earlier I had teamed up with Rick McIntyre to formally report Rick's observation of multiple scent-marking in a free-ranging pack in Yellowstone. We wrote:

> Most wolf packs are comprised of a pair of parents and their offspring of the past 1 to 3 litters such that the only pack members that are sexually mature are the parents (Mech and Boitani 2003). Thus scent-marking in those packs is restricted to the breeding pair. Nevertheless, some wolf packs are comprised of multiple same-sex individuals (Harrington et al. 1982; Mech et al. 1998; Stahler et al. 2013). In at least one study of captive packs with multiple same-sex mature individuals, more than one of those individuals scent-marked (Asa et al. 1990).
>
> However, we found no record of a free-ranging wolf pack with multiple, same-sex wolves that scent-marked nor mention of that possibility in the latest review of wolf scent-marking (Harrington and Asa 2003) (McIntyre and Mech 2024).

I had forgotten that I had seen the Ellesmere pack multiple-mark even before reporting it for captives (Asa et al. 1990).

As for territorial marking, there wasn't often a chance for me to gather much information about that because Ellesmere wolf-pack territories are much larger than those in most areas, primarily because the prey density is so low (discussed later). Generally, wolves establish their dens toward the center of their territory (chapter 11), so I rarely had the opportunity to observe territorial interactions, which usually happen along the territorial boundaries (Mech 1994a).

However, something unusual happened in 1992, and I never learned as much about it as I wanted. It seemed as though some wolf adversaries to the Eureka pack must have taken up residence within about 5 miles (8 kilometers) of that pack's usual den. It was the only year I observed my wolves chase or attack outsiders, and they did so on three occasions (Mech 1993a). It was also the year when I observed the most scent-marking, and when the wolves were more alert and observant as they marked and scratched (Mech 2006).

That year the Eureka wolves also marked the easternmost locations where they traveled, suggesting a territory boundary. The wolves often seemed to be in an unusual state of arousal and aggressiveness. And that is also the year when a companion and I did a howling test some 5 miles (8 kilometers) east of the den and three strange wolves showed up from the east (see above). These observations all seem to evince that a new pack was trying to establish itself, or already had, not far from the den of the wolves I was working with.

A related aspect of scent-marking is ground scratching, behavior by breeding wolves that has been studied more as an adjunct to scent-marking than as a separate behavior (Peters and Mech 1975). Wolves scratch the ground much like dogs, quite vigorously, using all four feet. Ground scratching is thought to leave scent from various glands on the feet as well as a visual sign (Peters and Mech 1975; Harrington and Asa 2003). Most studies indicate that, although wolves do not always scratch the ground along with scent-marking, they don't often scratch without it, and that the seasonal frequency and occasion of ground scratching parallels that of scent-marking.

Mech (2006) recorded Ellesmere wolves ground scratching done (1) with no other marking behavior; (2) followed by FL urinations or RL urinations; (3) preceded by an FL urination, RL urination, or squat urination; (4) preceded and followed by FL urination; or (5) preceded by defecation. These observations are similar to the findings of Peters and Mech (1975) for wolves primarily in winter, and by Mertl-Millhollen, Goodmann, and Klinghammer (1986) with captive wolves.

During the Ellesmere study, either male or female breeding wolves, or both together, scratched the ground at times throughout the summer study. Of 16 summers when I observed mated wolves, I recorded no scratching by them during seven, and no more than three scratching events during six (Mech 2006). Only in 1992, 1993, and 1996 did I observe wolves do much scratching, and both breeders did so.

During those years, it was the breeding pair, the same individuals each year since 1989, that scratched regularly, but I saw little scratching during 1989–1991, 1994, or 1995. In 1992, this pair scratched and marked the most, primarily (1) near a muskox they had killed and at which a stranger wolf had also been feeding a few minutes before, (2) near the weather-station landfill where at least one nonpack wolf had been feeding regularly, and (3) along the easternmost location (territory boundary?) where I saw them that summer (Mech 2006). Whitey scratched at several places where she FL urinated and at one place she did six separate bouts of scratching. She did much scratching that year, with several scratches per bout, much more than the male.

I observed an excellent example of the context and significance of territorial scent-marking and scratching on July 23, 1992. Whitey and/or her mate had killed an old bull muskox about July 12 and had been feeding on it since. With food being so scattered on Ellesmere, and other wolves being in the area, it probably was inevitable that wolves other than Whitey and her mate would find the kill and try to feed on it.

Sure enough, that is what happened. According to my notes, starting from when the pair was still at the den:

> 0446 hrs. The male started howling and Whitey and pups joined and howled 32 X and both adults headed NE. We followed and . . . intercepted the 2 wolves ~ 1 mile S of camp.
>
> We then headed to the muskox kill and watched as the 2 wolves approached it from over the hill to the SW. They did not come straight to the kill. Instead, they spent ~20–30 minutes checking all the scent marks around the area, especially 100' away and they did more marking and scratching and especially scratching. Both wolves were very intent and excited and looked around a lot and circled the area from W to N to E and then fed on the kill.
>
> They then headed E uphill, looked NW and both started running NW across creek. I zoomed up E hill and saw them heading toward a third wolf running down valley Eastward ~1–1.5 mile away. We tried to follow but only saw our 2 wolves ~2–3 miles [3.2–4.8 kilometers] NW of hill heading E.
>
> We returned to m.o. [muskox] kill and found our 2 wolves there feeding. Whitey also did much more checking scent-marks and scratching and marking. Both RLU'd on m.o. head while feeding on it, several times.

From the circumstances during which wolves scratch the ground, it would certainly seem that ground scratching reflects an increased level of aggressiveness. I thought the wolves I watched scratching the ground seemed hyper alert or upset. What function this scratching serves in that regard is still unknown. As mentioned above, it would spread scent from the feet and be a visual signal, and maybe that is all. However, Paquet (1991) believes scratching could either enhance a wolf urine-mark or perhaps alter the meaning of an existing one.

Scent-marking, scratching, and/or howling are very important for wolves because they are so competitive and must continually assert their presence against potential enemies, primarily other wolves. The creatures generally den and raise their pups toward the center of their territory as far as possible from neighboring packs (chapter 11). Pups are extremely vulnerable during their first few weeks, and the competitiveness of wolf packs extends even to the point of killing their neighbor's newborn offspring.

This extreme behavior (infanticide) was little known for much of the period while I was studying the Ellesmere wolves. Two articles had suggested it (McLeod 1990; Latham and Boutin 2011), and it had happened in Yellowstone a few times, the first in 1996. However, little attention had been paid to it, and it had escaped my notice until details about it were published in 2015 (Douglas Smith et al. 2015).

The original den that I worked with on Ellesmere seemed to be safe from whatever neighboring packs there might be, and for the first few summers there I saw nothing that made me think any wolves other than the ones I was observing were anywhere nearby. That is, until 1990, when the pack consisted of a single pup and only three adults. Conceivably, other 1989 pack members remained in the area, but if so, I never saw them near the den.

At least two or three, and possibly four, new wolves did appear, however, about a half kilometer from the den on July 22, and both Whitey and Mom individually chased them away. The situation was such that I could not see the details of the chase or the end result. The only other information I had was that within about two hours, all three adults were back at the den sleeping. Were these strange wolves some of Mom's 1988 or 1989 offspring? If so, they certainly were not welcome anywhere near the den. Chances were good that these were unrelated, or at least less related, neighboring wolves.

I had no idea how large the Eureka pack's territory might be or where its boundaries might lie. The distant places I had tracked that pack to by ATV or air by 1990 were many kilometers from the den, so I assumed that territories there were huge and that neighbors would be of little consequence to my studies around the den. Still, where were these strange wolves from, and how far did they travel to get as close to the den as they did? Did wolf territoriality work differently here than in Minnesota and Denali where I had studied it? True, I had seen occasional trespasses in those other areas, so maybe the 1990 strays were just out trespassing.

It turned out that the 1990 appearance of strangers was a portent, for 1991 held some more surprises along these lines. First of all, when I arrived in June 1991, weather-station folks reported just having seen a pack of seven wolves in the area the day before. The very next day, I spotted a group of four near the weather station. I checked them by getting close to them and trying to see how they responded to me. They appeared more afraid than interested or even tolerant. I concluded these were not any of the wolves I had worked with in previous years.

Three weeks later, I ran into four wolves again several kilometers away, but they did not seem to be the same group. They responded differently to me than the other four, and two seemed to recognize my signals. These four followed some of the same routes around the area, namely, taking shortcuts across topographic features and so forth, as did the wolves in 1989. Thus, I felt that at least some members of this pack were those I had worked with that year that had at times attended the wolves' 1989 rendezvous site. The question remained, however, as to why they were not now helping Whitey and Mom like in 1989. I never did figure out which they might be.

The next sighting of a strange wolf was far more dramatic. Just a few days after I had seen the second group of four, weather-station personnel reported that they had seen Whitey, Mom, and Blackmask chase a strange wolf into a gulley near the area's runway and had come back all bloody. I headed right there and found a badly wounded wolf lying in the mud of a bank seep. It was covered with blood around its head and rump and could barely arise. It struggled up and walked slowly, was lame and rested a lot, and I was pretty sure it would soon die. When I got back to it several days later, all I found was the wolf's well scattered remains, indicating it had been eaten by the wolves.

That altercation took place in an area most likely to host overlap of wolf packs, Eureka, with its human occupants. Although year-round, sometimes fewer than 10 people inhabit the area, from about April through July, there may be several times that. Some of those folks throw food to the wolves or leave food for them (chapter 1), and before an incinerator was installed in the 1990s, garbage accumulated in the landfill. Thus, probably all the wolves within many kilometers would sooner or later have learned to visit the place to sample some free food. When Morgan and Dan GPS radio-collared other packs in the region during 2014–2018, several did indeed still use the Eureka area, which confirmed this suspicion.

The trend of strange wolves showing up continued in 1992 and actually peaked then. As soon as I arrived at Eureka, weather-station personnel told me that near a summer research camp 22–25 miles (35–40 kilometers) east of the Eureka pack's den, there was another pack denning. That area is not far from where the Eureka pack had taken its single pup on its long trek in 1990.

I first encountered unknown wolves directly in 1992 while running the July 15 test mentioned earlier to see how far wolves could hear my howling. That's when three wolves showed up unexpectedly behind me looking around for the strange (human) howlers.

Those could also be some of the wolves I spotted two days later right near the weather station. They did not appear to be the same wolves that surprised me two days earlier, but I could not be certain. One did come up to me readily, and the other two hovered around. The three eventually headed along the old road paralleling the shore of the fiord toward the fiord den, where Whitey had moved her pups. Then one veered off, and I followed the other two. At 0037 hrs, one suddenly stopped, dug up a cache, and pulled out eight large pieces of meat; presumably having cached it there himself.

I had the impression that these two wolves and their former associate were immature and that those seen yesterday were more mature. These, for example, seemed more carefree and frivolous, whereas yesterday's three seemed more serious and intent. On the other hand, in both groups, there seemed to be one slimmer wolf, and only one of the others each time came over to me readily. A wolf's attitude can change quickly. So maybe they were the same three. I continued to follow the two, knowing they would soon pass below the fiord den where the pair was headquartered.

They did just that, heading along below the den, which was up the hill just ~300 meters but showed no sign of hesitation or fear or any indication they knew other wolves were in the area. When about 200 meters west of the area just below the den, one of the wolves veered off the road and jumped a hare. The hare ran straight downhill toward the fiord ~50 meters away, with both wolves in pursuit. There was some zigzagging, but most of the chase was straight, and I could see the wolves were going to catch the hare at the shore.

At 0055 hrs a blood-curdling shriek emanated from the shoreline, and each wolf looked up with half a hare in its mouth.

Instantly both wolves then shot off running, and from behind me down the hill burst Whitey and her mate who—no doubt attracted by the squealing hare—spotted the two intruders and darted toward them. I watched as Whitey and her mate chased the other two for a kilometer up a hill and out of site. If only I could have followed them.

Not until 0205 did I find the pair back at the den. They were not bloody at all so may not have ever caught up with their quarry.

Given the proximity of all these intruding wolves in 1992, it is no wonder then that Whitey and her mate did the most scent-marking and especially scratching that year, which seems to reflect a high level of aggressiveness. The July 23 observation detailed earlier featuring Whitey and her mate scent-marking and scratching vigorously around the muskox carcass and then chasing a wolf that almost certainly had been feeding there just before they arrived attested to that.

I actually got to directly witness some of that aggressiveness close-up several days later, when on August 4 I was following the whole pack, pups included, as they approached the weather-station landfill (Mech 1993a). My field notes tell the story:

> Our male had been approaching intently and alertly and slowly and saw the strange wolf when ≥100′ away, just about when we did.
>
> Instead of rushing the wolf, the male walked toward it as though familiar with it. Meanwhile, the stranger had already detected the male and had assumed a sitting up, ears-back, muzzle-up, defensive grin pose.
>
> The male stood stiffly to the right of the stranger, but tail was not vertical. Rather it was only half horizontal. The 2 wolves stood like this for ~2 min. The scene was very much like that in "White Wolf" video after the musk-ox kill, where the subordinate wolves, sitting up subordinately approach the alpha male, and he pins them.
>
> I kept my machine running, so that I could immediately follow any chase, so we could not detect any growling. Each wolf was very deliberate in its stance, so that the whole interaction was very ritualistic.
>
> The 2 wolves moved slowly and the male was facing directly away from us, so I could not see his facial features. The other was sideways and several times had its lips retracted and teeth bared in a defensive grin [Fig 21.1] and did some defensive snapping at the male's back. It was like each was waiting for the fight to begin, but was reluctant to start it.
>
> Suddenly, the fight began, and I cannot now recall who started it, if I even saw that. In any case, the 2 wolves started to tangle after ~2 min [fig. 21.2].

Figure 21.1. Breeding male that I was studying in 1992 (right) encountering an intruding wolf, Eureka, Ellesmere Island, Nunavut, Canada.

Figure 21.2. Breeding male fighting intruder, Eureka, Ellesmere Island, Nunavut, Canada.

The strange wolf quickly took flight with the male chasing. Suddenly Whitey, who had by now arrived with the pups at N end of landfill behind us, came flying by joining the pursuit.

The wolves disappeared over the S edge of the landfill, and Tim [Tim Dalton, the English actor] and I headed there and watched the chase in the valley to the SW, into a large gully and out of sight.

After about 10 minutes, we saw a wolf pop up twice for a few seconds ~1/2 mile away in front of the fiord, and then Whitey returned across the valley to the landfill, and ~3 minutes later the male returned. Neither had blood or visible bite marks.

Total distance of the chase was estimated at 3/4 mile. When the male returned to Whitey and the pups there was no elaborate greeting. . . .

Thus, it seemed pretty clear both in summers 1991 and 1992 that neighboring wolves were living much closer to the den that year than in previous ones and that the denning wolves that I was watching were pretty wary of them. Plain old competition and territoriality easily explained the behavior I was seeing. That all happened before the Yellowstone wolf reintroduction and thus before anyone even knew that wolf packs tended to kill the neighbors' newborn pups (Douglas Smith et al. 2015).

It was also long before I would ever learn what the territorial structure of the Fosheim Peninsula's wolf packs was. In the summers of the early 1990s, it seemed like the territories must be relatively small because of the frequency of intruding wolves into the area that the pack I was studying used. This notion contrasted with my earlier estimate that this pack occupied an area of about 1,000 miles^2 (2,500 kilometers^2) during 1986 and 1987 based on all the places where I had seen them (Mech 1988a).

Furthermore, information from the single wolf, Brutus, that Dean and I GPS-radio–collared in 2009 also suggested that the territories were huge. Brutus had been the breeding male in the Eureka pack since 2003. Between when collared on July 9, 2009, and October 18, 2009, he covered an area of 640 miles^2 (1,642 kilometers^2) (fig. 21.3). Brutus turned out to be a member of a pack of at least 20 wolves, including grown pups.

There is no way of knowing how many, if any, other wolves accompanied Brutus throughout his fall and winter journeys, but each of seven times he was seen between August 7, 2009, and April 16, 2010, around the weather station or den he was part of a group of 8 to 28 adults (Mech and Cluff 2011). Furthermore, the rate at which Brutus's location clusters indicated large kills also indicated that he was traveling with other wolves (Cluff and Mech 2023).

This contrast between what I was observing with the wolves around the den and the information from Brutus's travels when collared plagued me. The pack-size difference did as well, that is, the most adults I had ever seen associated with a den was eight, and the average four. However, I would often hear from the weather-station workers that during winter they were seeing packs several times that size. Those folks thought such large packs might live on other islands and could only be coming to Eureka from nearby Axel Heiberg after the sea had frozen.

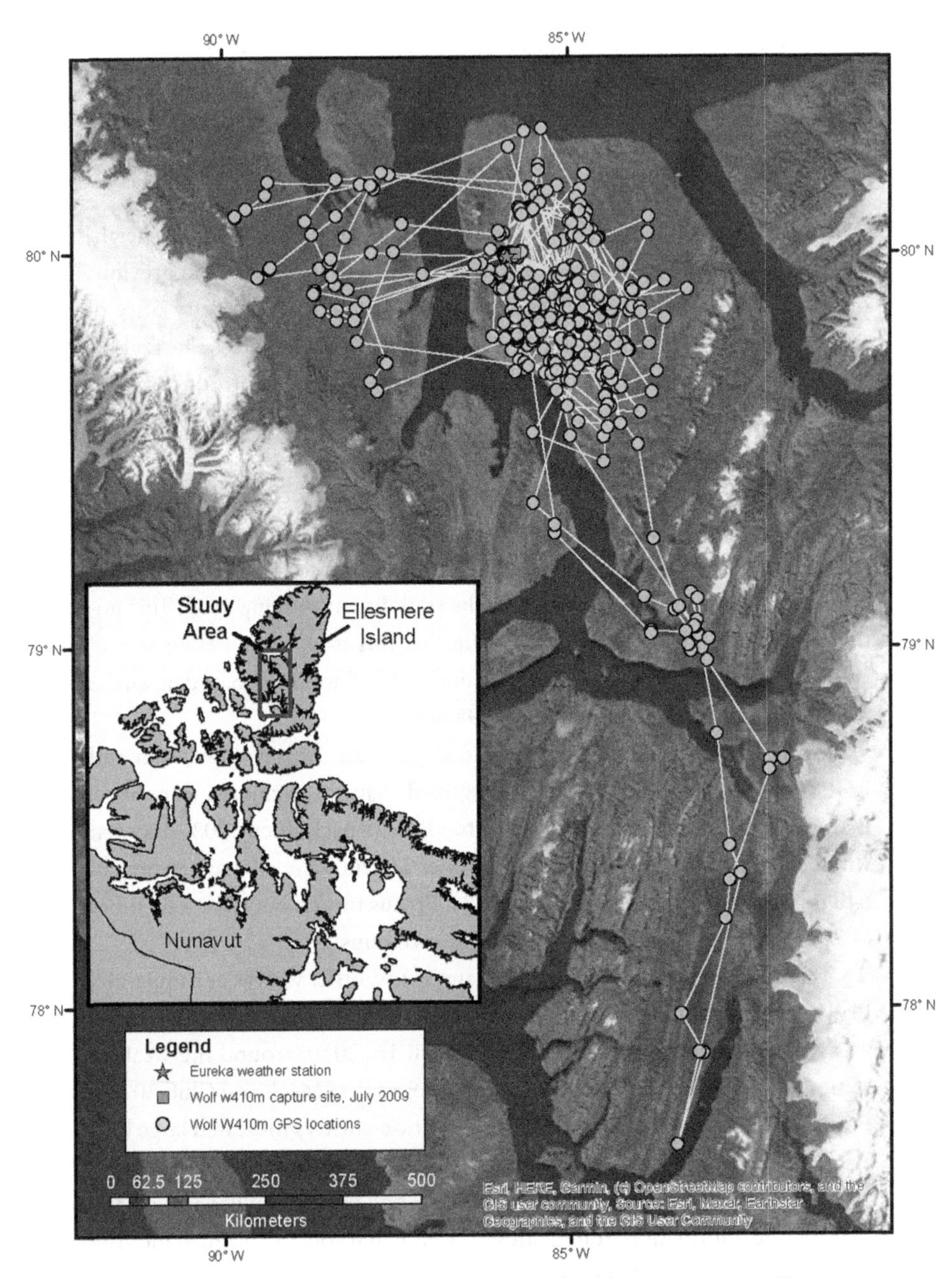

Figure 21.3. GPS locations of Brutus and his pack of at least 20 wolves on Ellesmere Island, July 9, 2009, through April 10, 2010. From Cluff and Mech 2023. The area Brutus used from April 10 to October 18 was about 1–2% of the mass of points near the top of the map. The legend and scale pertain to the inset map.

Figure 21.4. From 2014 through 2018, Morgan Anderson (pictured) and colleagues Dan MacNulty and Dean Cluff GPS radio-collared 10 wolves on Ellesmere Island, Nunavut, Canada. From Anderson et al. 2019.

When Brutus, who had been in the Eureka area for years, showed up in a much larger pack when collared, that finding exploded the Axel Heiberg theory. True, large packs from Axel might have been visiting Eureka, but Brutus showed that those reported over the years could just as well have been resident on Ellesmere itself. All this contrasting information made me hungry for additional data to help solve the issue, data that could only come from GPS radio collaring more wolves.

And the new data did finally come. During 2014 through 2018, my colleagues Dan MacNulty and Dean Cluff, with the addition of Morgan Anderson (fig. 21.4), managed to GPS radio-collar 10 wolves from six more packs on the Fosheim Peninsula and nearby Axel Heiberg Island (table I.4). Although these workers never did collar any wolves that used the original den that I worked with, they did study all the packs surrounding that pack's territory.

As it turned out, the territory of the pack I had long studied was one of the smallest in the area (fig. 21.5), and that pack included some of the fewest adult members as well. Thus the explanation for the apparent contrasting early findings was that (1) wolf-pack territories on the Fosheim Peninsula were highly

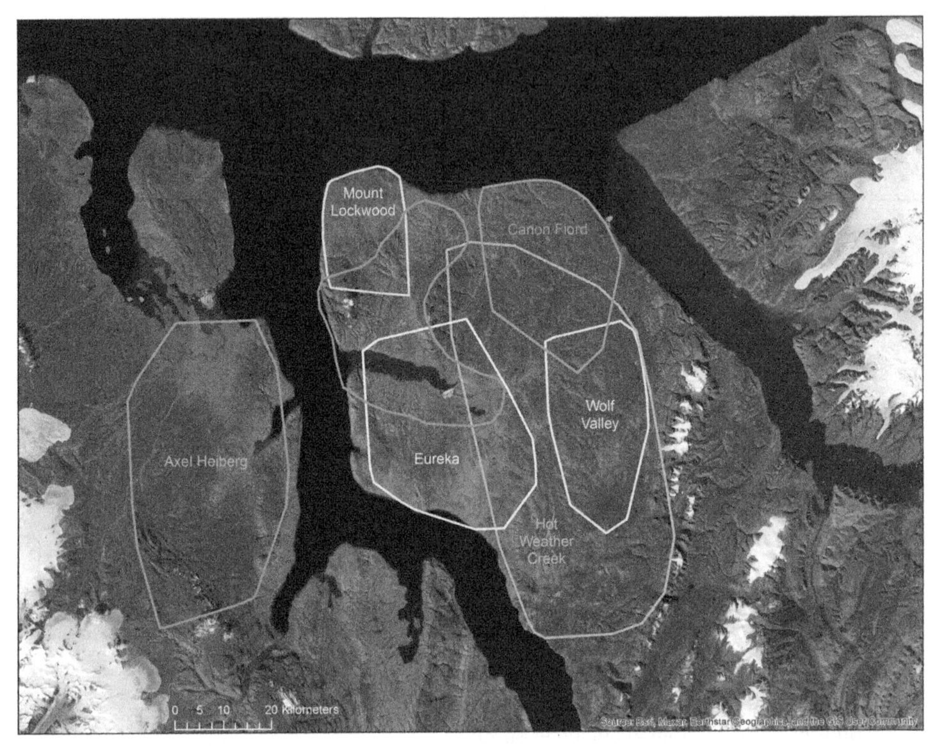

Figure 21.5. Wolf-pack territories on the Fosheim Peninsula, Ellesmere Island, Nunavut, Canada. The unnamed, irregular polygon in the center approximates the territory of the rock-cave wolves that Dave studied mostly north of Slidre Fiord (fig. 1.3).

variable in extent, with their summer areas ranging from 156–1,043 miles^2 (400 to 2,670 kilometers^2) (Anderson et al. 2019; table I.4); (2) the number of summer adult pack members was highly variable as well, from 2 to 16; (3) some packs apparently had multiple breeding females; and (4) Brutus might have shifted his pack affiliation during some years, or at least his pack started denning in a whole new area.

Probably the reason weather-station personnel often saw larger packs during winter than I worked with nearby in summer was that once Slidre Fiord froze over, the pack directly south and across the fiord from the station was able easily to cross the ice. This shortcut would have saved those wolves a few hours' worth of travel. In 2016, for example, that pack, which is the one whose den Brutus attended in 2009, consisted of 10 adults and 11 pups. A similar situation no doubt pertained in 2009–2010 when Dean and I studied Brutus's movements by GPS collar. Soon after we collared Brutus, we saw him with 11 other adults and at least 3 pups. There was even some evidence during 2009 that Brutus was associated with two dens (chapter 9).

This explanation for the large winter packs does not rule other possibilities. It could still be the case that large winter packs sometimes could result from neighboring packs that are independent, though probably related, during summer tolerating each other in some areas during winter.

Brutus, and presumably his pack, besides using an area of 640 miles^2 (1,642 kilometers^2) up until October 18, 2009, on Ellesmere's Fosheim Peninsula, then crossed some 9 miles (13 kilometers) of newly frozen fiord just west of Ellesmere to Axel Heiberg Island and back several times, covering an area 41 x 34 miles (65 by 55 kilometers) on Axel's east-central lobe. They later made another round trip to another part of Axel several kilometers to the southeast in early March. The Axel Heiberg Island locations added 765 miles^2 (1,960 kilometers^2) to their range which then totaled 2,594 miles^2 (6,640 kilometers^2), more than half the size of Connecticut (fig. 21.3).

During January 19–29, 2010, Brutus at least, and probably his whole pack, made a 164 mile (263 kilometer) straight-line distance foray southeastward to the vicinity of Bauman Fiord (fig. 21.3), returning from January 29 to February 1 (Mech and Cluff 2011).

Although Brutus's pack's summer territories (n = 3) averaged 550 miles^2 (1,405 kilometers^2) in extent (table I.4), the pack covered an area four times that size in winter and might have trespassed on several other pack territories (fig. 21.5). Perhaps it was able to do that because of its size. In Yellowstone National Park, larger packs tend to win altercations with other packs they might tangle with (Cassidy et al. 2015). Brutus's pack was the largest in the area, based on data from a few years later (Anderson et al. 2019) and on 34 observations of Ellesmere packs from 1967 to 1991 (Miller and Reintjes 1995).

It's also possible, however, that given the large areas being covered, the wolves might not be around to mount significant defense of their territory, or territory defense is lower priority than just finding muskoxen in that vast area.

I knew none of this in the early 1990s when the pack living north of Slidre Fiord was having to fend off so many intruders. However, after my colleagues collared wolves from several other nearby packs, it became clear that that pack's territory was surrounded by territories of other packs (fig. 21.5), much like the population structure in most other areas. I even learned later that another colleague who had assisted with documentaries of these packs had observed a case of infanticide and knew of another in those packs (Kira Cassidy, Yellowstone Center for Resources, unpublished data, 2023). Thus, it was not that unusual that my pack had to deal with strange wolves. It all finally made sense.

: : **22** : :

The Bigger Picture

With my long-term focus on the behavior of the wolf pack in the vicinity of Eureka, I myself only found hints about the ecology of the population that my study pack was part of on the Fosheim Peninsula. That broader information would have to come from my compatriots. In 2010, I had encouraged my former graduate student, Dan MacNulty to join me and Dean Cluff on Ellesmere to continue studying the wolves there, especially via outfitting more wolves with GPS radio collars (chapter 9).

Collaring more wolves would allow not just Dan, Dean but also me to follow their movements throughout the year from our offices merely by tuning in to a website. We then also could detect the clusters of locations that would indicate where the radio-collared wolf and any associates remained at specific points for longer periods and had fed on kills (Cluff and Mech 2023).

Like we did in 2010, my colleagues could then later check these location clusters from the ground. With more wolves collared, we could then estimate wolf kill rate and possibly predation rate. To find the wolves to collar would require checking various possible dens, which would yield other information such as pack size and pup counts and allow estimation of the wolf density. This information would provide a much more complete picture of the ecology of the wolves whose behavior I had studied for so many summers.

By 2014, the necessary funding, permits, and logistics had been arranged to allow my colleagues to return to Eureka and collar more wolves. The Government of Nunavut wanted to assess wolf predation on Peary caribou, so the timing was ideal. Working with the Department of Environment allowed the researchers to communicate better with the Inuit and the decision makers. The Nunavut area biologist Morgan Anderson had a background in wolf ecology, projects in the area already, and experience with aerial capture, so she joined the team. The Iviq Hunters and Trappers Association in Grise Fiord was looped into the work, and summer student Etuangat Akeeagok assisted in all aspects of the fieldwork.

Between 2014 and 2018, these researchers captured six female and four male wolves from six packs on the Fosheim Peninsula and nearby Axel Heiberg Island (table I.2). They weighed and measured them (table I.3; fig. I.1) and outfitted them with GPS radio collars (Anderson et al. 2019).

Five females weighed 68.9 pounds (31.3 kilograms) and four males were slightly larger at 81.6 pounds (37.1 kilograms), generally lighter than most wolves, which females average 77–88 pounds (35–40 kilograms) and males 95–99 (43–45 kilograms) (Mech 1970) but well within the range of those in the Northwest Territories/northern Alberta (Kelsall 1968; W. Fuller and Novakowski 1955) and Alaska (R. A. Rausch. personal communication, in Mech 1970).

The GPS collars of those wolves yielded 3,224 locations, and the team investigated 592 location clusters (≥2 locations within 50 meters of each other). There they found remains of 83 muskox kills, 2 scavenged muskoxen, 47 probable arctic-hare kills, 1 dead wolf, and 1 seal, which was probably scavenged (Anderson et al. 2019).

As mentioned above, the packs of the GPS radio-collared wolves ranged over summer territories of 156–1,043 miles^2 (400–2,670 kilometers^2) (fig. 21.5). Their summer pack sizes varied from 2 to 16 adults, with the number of pups per pack varying from 0 to 11 (table I.4). It is difficult to explain the relationship between the 2014–2018 territories of these packs and the locations and movements of Brutus in 2009–2010, for Brutus and his pack used parts or all of these other packs' territories (compare figs. 21.3 and 21.5).

Possible explanations include the following: (1) Brutus's pack of 20+ dominated most of the Fosheim Peninsula but broke up after Brutus died in April 2010, and the remaining members of the pack formed several smaller packs; (2) as mentioned earlier, Brutus's pack of 20+ was so much larger than any other pack that it could travel wherever it wanted to with impunity because of its size (Cassidy et al. 2015), so merely trespassed through all the other pack territories. The 2010 to 2014 gap in studies of these wolves makes it difficult to choose between these two possibilities. A third possibility is that Brutus was not always traveling with his pack. If that were the case, however, he would be far more vulnerable to attack by resident packs than if he were traveling with his own pack.

The summer wolf-population density during the 2014–2018 study was estimated at 6.4–16.6 adult wolves/1,000 miles^2 (2.5–8.0 adult wolves/1,000 kilometers^2) and 9.7–25.1 adults and pups/1,000 miles^2 (3.7–10.4 adults and pups/1,000 kilometers^2) (Anderson et al. 2025). Because of the impossibility of studying this population during winters, these summer densities represent the best estimates available, even though population densities for almost all other areas are winter densities. Thus, the Ellesmere wolf densities are not directly comparable to those from other study areas. Still, as expected, the Ellesmere wolf densities fit among the lowest densities anywhere (T. Fuller, Mech, and J. Fitts-Cochran 2003), especially given that summer densities are always higher than winter densities because of the intervening mortality.

The lowest wolf densities generally are found in the highest latitudes where prey biomass is lowest, and Ellesmere marks the highest latitude from which a wolf density has been reported. The next highest latitude with a known wolf density is northern Alaska (70–71° north), about 10° farther south than Eureka. There the winter density was about 18 wolves/1,000 miles^2 (7/1,000 kilometers^2) (Adams and Stephenson 1986). In contrast, one of the highest wolf densities occurs in Yellowstone, with about 128/1,000 miles^2 (50 wolves/1,000 kilometers^2) (Mech and Barber-Meyer 2015). There, the amount of prey was about 12,800 (5,000 elk per 1,000 kilometers^2).

On the Fosheim Peninsula, the amount of prey was about 770–1,570 muskoxen/1,000 miles^2 (300–614/1,000 kilometers^2) (Fredlund et al. 2019), which easily explains the difference in wolf density between Yellowstone and Ellesmere (T. Fuller, Mech, and J. Fitts-Cochran 2003). To estimate how many of the available muskoxen the Ellesmere wolves killed each year, Anderson et al. (2025) were able to check wolf-location clusters for the Eureka pack, which numbered four adults and two pups. From July 2017 through June 2018, the pack fed on 40 muskoxen kills and scavenged two. That number would amount to about a muskox each 8.5 days. Rough calculations and assumptions about muskox weight, edible tissue, and wolf-food requirements led the workers to estimate that each member of the Eureka pack would also have to have killed an equivalent of about 115–228 adult arctic hares per year as well (Anderson et al. 2025).

When Anderson et al. (2025) made these estimates, the Eureka pack was living primarily south of Slidre Fiord in the area it had moved to during the last few years of my investigation rather than where I had spent most of my summers. Nevertheless, this ecological assessment provides a general picture of the relationship between the overall wolf and prey populations in the general area.

: : **23** : :

Climate Change

The perils of climate change are upon us. As with most scientists, I was introduced to the problem long ago. In my case, it was 1988 while I was part of a panel consulting for Glacier National Park on dealing with future wolf issues. The park biologist introduced the subject, known then as "global warming," as background because he knew that such a major pervasive influence could eventually overwhelm any advice the panel could give or decisions the members might make.

I have often been asked about the possible effect of climate change on wolves. My pat answer has usually been that the wolf as a species is adapted to every climate north of about 10° north latitude and feeds on every kind of animal at that full range of latitudes. The species is highly adaptable, with its feast-or-famine type of feeding, its fecundity, and its ability to travel long distances in short periods. Therefore, the species should be able to tolerate whichever way the climate changes in the northern hemisphere. In other words, although like most every other creature, the wolf will be affected by climate change, as a species the wolf does not appear to be in danger from it. Wolves live over most of the northern hemisphere, so any force that would endanger them would have to affect much of the whole northern half of the Earth.

As mentioned earlier, arctic wolves possess traits adaptive to their environment, being only one "ecotype" of wolves that have been identified based on slight differences in their genomes (Schweizer, Robinson, et al. 2016; Schweizer, vonHoldt, et al. 2016; Sinding et al. 2018). Thus, one might argue that it could be possible that although the wolf as a species might be resistant to climate change, some specific wolf ecotypes might be more susceptible to it.

Schweizer, vonHoldt, et al. (2016) did indicate concern that the wolves of the Arctic and High Arctic might be endangered by climate change, citing Mech (2004). However, as detailed later, I (Mech 2004, 92) did not claim that the Ellesmere wolves were being endangered by climate change, but only that they might have been affected by it. True, I did indicate that the changes I documented could be "dramatic and devastating to the wolf-prey ecosystem." Still, those features mentioned above should enable wolves to withstand not just general climate changes but also local and/or temporary climate-change effects. The wolves affected by the changes I (Mech 2004) documented were producing pups again within a few years.

There is no question that wolves' primary lifestyle—that is, their predatory practices—are often affected by weather. For example, snow accumulation greatly affects wolf predation via its influence on prey nutritional condition (Mech, Frenzel, and Karns 1971; Peterson and Allen 1974; Mech and Peterson 2003; Wilmers et al. 2020). Drought is another such condition (Wilmers et al. 2013).

And on Ellesmere, I found that the arrival of early winters (Mech 2001) during 1997 and 2000 was the most likely cause of both the starved muskoxen carcasses I discovered and the lack of reproduction in both muskoxen and arctic hares during some years. My explanation follows: "Herbivores gain most of their nutrition during summer when vegetation grows, and they depend on the fat they store during the summer-replenishment period to carry them through the following winter. In this study area, this usual replenishment period only lasts from about 1 July, when winter snow has disappeared and vegetation sprouts new growth, to about October 1 when snow covers the vegetation. During 1997 and 2000, however, the summer-replenishment period was cut almost in half" (Mech 2004, 90).

The early onset of winter conditions in summer would have ended the replenishment period in mid-August, reducing it by about 50%. Not only would these conditions foster prey starvation, they would also lead to poor reproduction and fewer quarry the next year. Further, because wolves tend to prey disproportionately on young-of-the-year, the canids' hunting success would no doubt have been reduced. That problem would be especially critical during summer when it is so crucial to fortify growing pups. Those issues could easily explain the lack of wolf reproduction that I found during the following years.

Table 23.1. Weather parameters in study area during months preceding winters in 1997–1998 and 2000–2001 compared to long-term norms (Environment Canada)

	Daily mean temperature (°C)			Monthly precipitation (mm)			Monthly snowfall (cm)			Month-end snow cover (cm)		
Month	**Norm.[a]**	**1997**	**2000**	**Norm.[a]**	**1997**	**2000**	**Norm.[b]**	**1997**	**2000**	**Norm.[b]**	**1997**	**2000**
August	2.9	0.7	0.4	11.8	25.6	33.8	4.0	12.6	23.8	0.0	4.0	3.0
September	-8.3	-10.4	-10.2	9.7	15.6	10.0	10.9	23.4	10.0	6.0	13.0	9.0

Note: Data are presented for 1997 and 2000 because the hypothesis is that those aberrant summers caused demographic changes a year later (Mech 2004). For the rest of the winter period in each of these two years, these parameters were normal or more favorable.

[a] 1961–1990.

[b] 1947–1990.

Was the onset of early winters a result of climate change? Quite possibly. So far north, small differences in temperature and moisture like I documented (table 23.1) can mean major differences in ground conditions, such as snow cover or icing that not only affects plants but also hinders herbivore access to the plants. Although a major result of climate change is an increase in temperature, rather than the decrease that Mech (2004) found, increases in extreme conditions can also be caused by climate change (Karl, Nicholls, and Gregory 1997; Karl 1999; Gitay et al. 2002).

The high precipitation and low midsummer temperatures during 1997 and 2000 as well as the resulting early snow were the most extreme in the five decades since weather records for the area were first kept.

The main cause of muskox die-offs in the High Arctic is freezing rain, rain-on-snow or other ground-fast icing events. Forage gets locked beneath impenetrable ice, sometimes for the entire winter when these events occur in autumn. Unable to access sufficient forage, muskox (and Peary caribou) starve in large numbers, if icing is widespread and they cannot move to accessible forage. Several of these events in the 1990s and early 2000s have been recorded, with mortality over 90% (Miller and Barry 2009; Miller and Gunn 2003).

The Eureka area is protected from storm systems moving from the west by the Princess Margaret Range on Axel Heiberg, and icing events have not been common there. However, there is significant uncertainty about how climate change will shift precipitation patterns, and an increase in frequency or severity of icing could have catastrophic impacts on the wolves' prey.

Figure 23.1. I could not help but ponder how changes in the timing and amount of sea ice could affect Ellesmere's weather and ecology. Photo courtesy of Dean Cluff.

Changes in timing and amount of sea ice could also affect the weather and ecology of the area in unpredictable ways (fig. 23.1). As sea ice lessens, wolf dispersal and possible outbreeding from wolves farther south would be reduced, possibly adding to inbreeding depression in the wolves.

If any wolves would be impacted by climate change, directly or through their prey, it would be those in the High Arctic including Ellesmere. The Arctic is warming almost three (World Wildlife Fund 2023) or four times faster than much of the rest of the world (Rantanen et al. 2022). From 1972 to 2007 the Eureka weather-station measured a 3.20°C increase in annual mean temperature (Lesins, Duck and Drummond 2010). Each decade, the maximum ice thickness decreased 1.4 to 2.0 inches (3.6 to 5.1 centimeters (±1.7 centimeters) from the late 1950s to 2016 (Howell et al. 2016). Current estimates project that the Arctic Ocean could become ice free in summer as early as 2030 (Kim et al. 2023).

Even without measurements, the changes were obvious to me over the 25-year period that I flew over the Arctic Ocean and explored the Ellesmere land mass. Eventually the familiar areas of the ocean that during my early trips were well frozen over began opening up, first in streaks and spots but in later years being wide open. On land in later years, I began to notice more and more "slumps," large swaths of steep banks whose underlying permafrost had melted, allowing them to slide downhill, exposing their raw, black former moorings (Ward Jones and Pollard 2021).

Although those kinds of changes, on Ellesmere (Dauginis and Brown 2020) or elsewhere, would have no direct effect on wolves, they could upset the more usual relationships of wolf prey to their environment discussed above. They could also represent the most apparent manifestations of more basic and dangerous changes like climate-pattern changes (Eley 2014) already seen elsewhere (World Wildlife Fund 2023).

Other changes in vegetation growth related to changing temperature and precipitation can also affect prey, and such changes have already been discovered on arctic tundra. “Satellites provide unequivocal evidence of widespread tundra greening, but extreme events and other drivers of local-scale ‘browning’ have also become more frequent, highlighting regional variability as an increasing component of Arctic change,” wrote Frost et al. (2019, 1). Thus, both positive and negative effects on prey-food acquisition are possible. Furthermore, there are no doubt other types of ecosystem effects we have yet to discover (Post et al. 2023).

Then there is the problem of inbreeding depression, with Ellesmere’s wolves being so close to the edge of the species’ distribution (Frévol et al. 2023). Longer seasons of open water south of Ellesmere even before loss of sea ice in the “last ice area” of the High Arctic could minimize immigration from the mainland for outbreeding.

A more insidious potential threat to the wolf population that my colleagues and I studied could arise from an increase in human accessibility to Ellesmere and associated islands. As seas remain open longer, human access will increase for exploration, mining (minerals and even coal), recreation, and other invasive activities. Presumably regulations would protect direct threats to all local wildlife, but development itself could usurp habitat and bring indirect threats including possible habituation of wolves to humans and thus conflicts. Still, most such activities would generally occur locally and not far from shore. The vastness of the interior of the area ensures that wolves and their prey will continue to thrive throughout the region despite much increased human incursion.

With the wolf’s mobility and its adaptability for various prey, most negative effects of climate changes would primarily be local and temporary. As a species, the wolf should survive climate change. Even the Ellesmere wolves that my colleagues and I studied.

: : **24** : :

Forever Wild

Yes, wolves as a species should survive climate change and the perils of the Anthropocene. They should also be able to survive many other kinds of changes that the future might bring. After all, they have survived so far through all the major disruptions wrought by humankind such as its mass dispersal across the world, the Industrial Revolution, and, so far, the Information Age.

One might question whether wolves can continue to survive their conflicts with humans, which mostly amount to depredations on domestic animals and pets or on key big game animals like moose and elk that human hunters treasure. To many rural folks across most parts of the wolf's worldwide, northern-hemispheric distribution, wolves are still the enemy (Boitani 2003). Furthermore, humans have not yet learned how to live with or prevent those conflicts so generally resort to lethal solutions, given that other methods succeed only temporarily or locally (Fritts et al. 2003).

These sorts of issues were always at the fore during the vast majority of the year when I was away from my Ellesmere "busman's holiday." True, wolves were faring well in much of the rest of the world. They were increasing in Minnesota, Wisconsin, and Michigan, and had been reintroduced to Yellowstone National Park and central Idaho and were spreading westward from there. In western Europe, they were also beginning to flourish. Throughout Canada and Alaska, wolves were continuing to maintain their numbers.

Yellowstone's wolf restoration, while biologically a great success, brought with it an ongoing controversy about whether the reintroduced Canadian wolves, considered "foreign wolves" by some wolf adversaries, were killing off all the park's elk or leaving the park and preying on nearby livestock. In Denali, some key park wolves were annually being trapped out in a critical corridor into the park where they were legally unprotected. Minnesota wolves were going on and off the federal Endangered Species List.

Nevertheless, my long-term Minnesota study on wolves and deer continued routinely, fostered by my assistants, technicians, and graduate students. Two experienced biologists carried out my day-to-day Denali research while my usual time there came during our spring captures of caribou calves. All the while, I was also tending to my graduate students at the University of Minnesota as well as to my duties with the International Wolf Center and the IUCN Wolf Specialist Group. Not to mention the "care and feeding" of 24 authors of various chapters of our 2003 book, *Wolves: Behavior, Ecology and Conservation,* which Luigi Boitani and I were working on for 10 years.

In the midst of all this activity, my great weeks each summer on Ellesmere astraddle my ATV alongside my wolf buddies living out their lives like they had for eons were a veritable godsend. Hanging out with them around the den or following them around far away from all the controversies and issues with wolves in the rest of the world was such a relief. This work wasn't counting the number of wolves in a pack from an aircraft, placing a point on a map of a new wolf location, or finding a scat full of beaver hair. This was sitting among a pack of wolves while they broke out into a raucous howling chorus or riding along with the pack while they headed off on a muskox hunt.

Not only had I finally found a den I could study up close, but the discovery had paid off in great dividends each year in the form of a blizzard of unique experiences and a sea of new data points. In my simple quest to see a white wolf and to finally observe a den, I had stumbled into a Mecca of wolf information: tidbits like watching one pup pounce on another or seeing a breeding female look up as a jaeger's cries signaled the return of her mate; opportunities like watching wolves catch hares; prospects like digging up a fresh cache and weighing its contents.

This long Ellesmere wolf experience turned out to be both a personal adventure and a professional bonanza. From a personal standpoint what could possibly have been better than to have spent all these summers living with this charismatic carnivore that had pervaded my life and career? That had allowed me to gain insights into my study subject that I could not have ever gained in any other way, not by the latest GPS collars, camera traps, genomic analyses, physiological tests, or the fanciest functional magnetic-resonance-imaging experiments. And from a professional standpoint, how else could I have obtained such a wide variety of new bits of information like the nursing schedule of the breeding female, the ambushing by parent wolves of offspring-chased hares, the motivating role of howling to get a mate to hunt, or the importance of hares to pup survival? More than 30 scientific publications, several popular articles, and four books document all this and so much more.

Being immersed in, and intimately surrounded by, all of this natural wolf behavior each summer was a far cry from being constantly reminded about all the various controversies surrounding wolves in most other places in the world. It's not like my other research projects had to deal directly with wolf management and controversy, but I as a US government and international wolf authority was regularly faced with so many of these issues. Fortunately, these almost universal problems with wolves were not ones that plague the wolves of the High Arctic. Those that we studied live 250 miles (400 kilometers) from the nearest domesticated creature, the sled dogs of Grise Fiord, and thousands of kilometers from domestic livestock. Thus, there is no reason for anyone to kill the predators. The area is too far away and expensive for recreational hunters to reach; they can harvest arctic wolves some 1,100 miles (1,800 kilometers) farther south. The Inuit in Grise Fiord can legally take wolves, and some that far north where the ocean is frozen for 10 months of the year. Nevertheless, the Inuit rarely venture to the Eureka area.

Conceivably, some dispersing wolves from the Fosheim packs might pass through Grise Fiord and attack a dog. However, the chances are low, and I never heard of it happening during all the years I studied them. Later, collared wolf 445 did walk right through Grise in 2016. She did not attack a dog or otherwise conflict with the village, but the local hunters all happened to be attending a fishing derby elsewhere, or they might have killed her.

A decade or so before I started my Ellesmere studies, the Eureka Weather Station kept sled dogs, and wolves did interact with them and sometimes had to be killed (Grace 1976), but no dogs are kept there now. In spring 1991, a wolf did nip the butt of a weather-station cook who was carrying meat from the refrigerator warehouse to the main station and had to be killed to check for rabies, which turned out negative. Those types of incidents have been so few and far between that they pose no danger to the local wolf population.

No, there seems to be little reason, anthropogenic or natural, that the wolf population that my colleagues and I studied, the wolves of several books and

TV documentaries and numerous scientific articles, should be in danger for the foreseeable future. Nor should their descendants. Their relatives much farther south and in various parts of both North America and Eurasia will probably long continue to bear the brunt of human controversy. The Ellesmere wolves, however—way off there near the end of the Earth—should long be able to continue their lives as natural as can be, wild and free of human wrath.

Acknowledgments

We thank Eureka Weather Station, the Polar Continental Shelf Program, and the Government of Nunavut for logistical cooperation; the US Geological Survey and the US Fish and Wildlife Service, National Geographic Society, BBC, Gulo Films, John Downer Productions, Plimsoll Productions, and Silverback Production for funding; the Iviq Hunters and Trappers Association for support and Inuit *qaujimajatuqangit* contribution; and numerous colleagues and companions mentioned in the text for their assistance with observations. We also appreciate the reviews of an earlier draft of this manuscript by Rolf Peterson and Luigi Boitani, the review of our scientific names by Steve Fritts, and the advice of Cree Bradley on photo selection. The following programs were used to help create/recreate graphs and or maps: ESRI (2023), ArcGIS Pro 3.2, Redlands, CA: Environmental Systems Research Institute; R Core Team (2022), R: A Language and Environment for Statistical Computing, Vienna, Austria: R Foundation for Statistical Computing, https://www.r-project.org/. Any use of trade, firm, or product names is for descriptive purposes only and does not imply endorsement by the US government.

Appendix I

Appendix Table I.1. Dates Dave Mech and/or his colleagues were in the Ellesmere study area (excluding 2014–2018) and sizes of wolf pack studied

Year	Dates in field	Number of days	Adult wolves	Pups	Total
1986	July 5–August 1	27	7	6	13
1987	June 23–July 3	10	7	5	12
	July 30–August 10	11			
1988	June 20–August 4	45	4	4	8
1989	June 14–August 11	58	8	4	12
1990	June 20–August 9	50	3	1	4
1991	June 13–August 8	56	3	2	5
1992	July 2–August 2	31	2	3	5
1993	July 1–August 6	36	5	0	5
1994	June 30–July 26	26	4	1	5
1995	June 29–July 8	9	3	0	3
1996	June 25–August 1	37	2	2	4
1997	July 2–6	4	(3)	0	3
1998	July 1–12	11	2	0	2
1999	Weather station info	–	(2)	?	(2)
2000	June 20–July 7	17	2	0	2
2001	July 1–13	12	(3)	0	(3)
2002	July 1–5	10	(4)	0	(4)
2003	June 29–July 4	5	3	0	3
2004	July 9–16	7	3	4	7
2005	July 8–20	12	6	3	9
2006	July 1–13	12	7	5	12
2007	June 30–July 5	5	(5) 3	(4)	(9)
2008	July 2–18	16	8	?	?
2009	July 3–17	14	12	3	15
2010	July 4–17	13	6	9	15

Note: Numbers in parentheses indicate reports from weather station personnel for May–July.

Appendix Table I.2. Radio-collared wolves, pack affiliations, and collar performance for wolf packs

Wolf	Pack	Sex	Age at capture	Transmitting days	Transmitting dates
W440	Axel Heiberg	M	2 yr.	384	July 15, 2014–August 3, 2015
W441	Cañon Fiord	F	2 yr.	164	June 30, 2014–December 10, 2014
W442	Wolf Valley	F	Adult	23	September 6–28, 2014
W443	Hot Weather Creek	M	Adult	725	September 6, 2014–August 31, 2016
W444	Eureka, Axel Heiberg	M	Adult	588	June 3, 2015–July 11, 2017
W445	Axel Heiberg	F	Adult	392	June 5, 2015–July 1, 2016
W446	Mount Lockwood	F	Adult	176	July 6, 2016–December 29, 2016
W447	Eureka	F	2 yr.	187	July 6, 2016–January 9, 2017
W448	Eureka	F	~6 yr.	350	July 15, 2017–June 30, 2018
W449	Eureka	M	~3 yr.	235	July 18, 2017–March 10, 2018

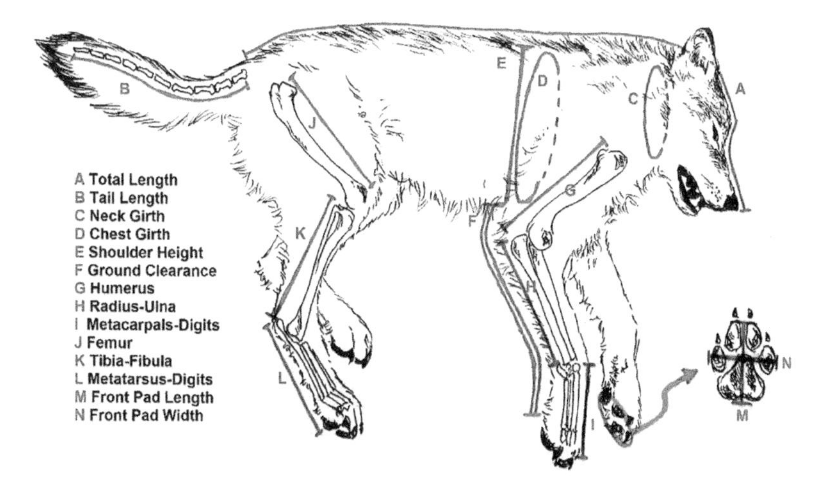

Figure I.1. Standard measurements of all wolves captured and collared on Ellesmere and Axel Heiberg Islands, 2014–2017. Measurements were made with a cloth tape.

Appendix Table I.3. Morphometric measurements for 10 wolves captured and fitted with radio collars

	Female			Male		
	Mean (cm)	SD (cm)	N	Mean (cm)	SD (cm)	N
Shoulder height	67.0	4.2	5	69.5	7.3	4
Chest girth	68.8	2.8	5	74.3	3.2	3
Body length	115.2	6.8	5	124.3	5.7	4
Ground clearance	40.2	1.9	5	41.5	3.7	4
Neck girth	41.2	3.4	5	45.7	2.9	3
Tail length	41.8	2.2	5	42.3	2.1	4
Humerus	24.8	2.1	4	26.5	3.5	2
Radius–ulna	25.5	2.6	4	29.0	1.4	2
Metacarpals–digits	19.0	0.8	4	22.5	9.2	2
Femur	29.8	1.3	4	30.5	0.7	2
Tibia–fibula	30.0	1.4	4	31.5	2.1	2
Tarsus–digits	24.3	3.1	4	25.0	2.8	2

Appendix Table I.4. Pack sizes and territory sizes for wolf packs monitored at dens and radio-collared, 2014–2018

Pack ID		2014	2015	2016	2017	2018	Mean
Eureka	Pack size	15–16	13 (3)	10 (11)	4 (3)	4 (0)	7.8
	95% MCP		1,652	1,429	1,135		1,405
Axel Heiberg	Pack size	7 (9)	9 (n/a)	6 (6)	6 (n/a)		7.0
	95% MCP	1,182	1,708	1,664			1,518
Hot Weather Creek	Pack size	2 (0)		2 (5)	2 (3)		2.0
	95% MCP	1,436	2,670	2,285			2,130
Cañon Fiord	Pack size	6 (3)		4 (n/a)		6 (4)	5.3
	95% MCP	822					823
Mount Lockwood	Pack size	5 (3)		2 (1)			3.5
	95% MCP			400			400
Wolf Valley	Pack size	4 (4)					4.0
	95% MCP	688					688
Rock Den[a]	Pack size				2 (5)		2.0
Vesle Fiord[a]	Pack size	3 (3)	5 (n/a)			4 (2)	4.0
Bay Fiord[a]	Pack size	2 (3)					2.0

Note: Parentheses indicate the number of pups observed. The mean pack size refers to adult wolves. Summer territory size is 95% minimum convex polygon (95% MCP) in km^2 (modified from Anderson et al. 2025).

[a] Bay Fiord and Vesle Fiord packs are included here but lived south of our study area and were never collared; wolves at the Rock Den, where Dave did most of his studies, had uncertain pack affiliation and were not collared.

Appendix Table I.5. Size of seasonal home ranges for 10 wolves (calculated with a north pole Lambert azimuthal equal area projection centered on the study area, latitude of origin 80°, central meridian -86°)

Wolf ID	Pack	Season	Area of 95% minimum convex polygon (MCP) (km^2)
W440	Axel Heiberg	Summer 2014	1,182.4
		Winter 2014[a]	3,587.5
		Summer 2015	1,707.6
W441	Cañon Fiord	Summer 2014	822.6
		Winter 2014	1,260.2
W442	Wolf Valley	Summer 2014	687.8
W443	Hot Weather Creek	Summer 2014	1,436.1
		Winter 2014	2,129.1
		Summer 2015	2,669.7
		Winter 2015	2,808.9
		Summer 2016	2,284.9
W444	Eureka	Summer 2015[b]	1,652.2
		Winter 2015[c]	7657.1
	Axel Heiberg	Summer 2016	1,664.3
		Winter 2016	5,286.9
W445	Axel Heiberg	Summer 2015	1,652.2
		Winter 2015[d]	5,805.5
W446	Mount Lockwood	Summer 2016	400.0
		Winter 2016	918.2
W447	Eureka	Summer 2016	1,429.0
		Winter 2016	1,032.3
W448	Eureka	Summer 2017	1,138.7
		Winter 2017	1,088.7
		Summer 2018	612.1
W449	Eureka	Summer 2017	1,131.0
		Winter 2017[e]	616.9

[a] Not including off-territory movements Feb. 13–15, Feb. 19–22, Mar. 12–15, Mar. 25–37; 6,027.0 km^2 for entire season.

[b] Starting June 17, after an off-territory movement following collaring.

[c] Not including off-territory movements Jan. 1–2 and Jan. 23–30; 11,297.9 km^2 for entire season.

[d] Not including off-territory movements Feb. 2–4, Mar. 14–18, and after Mar. 30; 14,304.1 km^2 for entire season until May 3 when this wolf dispersed off-island.

[e] Until Dec. 22, after which location fixes were intermittent.

Appendix II

Species Names

arctic fox (*Vulpes lagopus*)
arctic hare (*Lepus arcticus*)
arctic tern (*Sterna paradisaea*)
arctic willow (*Salix arctica*)
arctic wolf (*Canis lupus arctos*)
beaver (*Castor canadensis*)
beaver, prehistoric (*Dipoides* sp.)
beluga (*Delphinapterus leucas*)
bison (*Bison bison*)
black bear (*Ursus americanus*)
camel (*Camelus* sp.)
caribou (*Rangifer tarandus*)
collared lemming (*Dicrostonyx groenlandicus*)
coyote (*Canis latrans*)
elk (*Cervus canadensis*)
ermine (*Mustela erminea*)
flying lemur (*Cynocephalus volans*)
goose, brant (*Branta bernicla*)
grizzly bear (*Ursus arctos*)

gyrfalcon (*Falco rusticolus*)
jackal (*Canis aureus*)
leopard (*Panthera pardus*)
lion (*Panthera leo*)
long-tailed jaeger (*Stercorarius longicaudus*)
moose (*Alces alces*)
muskox (*Ovibos moschatus*)
narwhal (*Monodon monoceros*)
Peary caribou (*Rangifer tarandus pearyi*)
polar bear (*Ursus maritimus*)
ptarmigan (*Lagopus muta*)
purple saxifrage (*Saxifraga oppositifolia*)
red deer (*Cervus elaphus*)
red fox (*Vulpes vulpes*)
red knot (*Calidris canutus*)
roe deer (*Capreolus capreolus*)
seal, ringed (*Pusa hispida*)
snowy owl (*Bubo scandiacus*)
tapir (*Tapirus* sp.)
tiger (*Panthera tigris*)
walrus (*Odobenus rosmarus*)
wild boar (*Sus scrofa*)
wolf (*Canis lupus*)
wolverine (*Gulo gulo*)

Literature Cited

Adams, L. G., and R. O. Stephenson. 1986. *Wolf Survey, Gates of the Arctic National Park and Preserve—1986*. Anchorage, AK: US National Park Service.

Allen, J. J., M. Bekoff, and R. L. Crabtree. 1999. "An Observational Study of Coyote (*Canis latrans*) Scent-Marking and Territoriality in Yellowstone National Park." *Ethology* 105:289–302. https://doi.org/10.1046/j.1439-0310.1999.00397.x.

Anderson, M., D. MacNulty, H. D. Cluff, and L. D. Mech. 2019. *High Arctic Wolf Ecology: Final Report 2014–2018*. Igloolik, NU: Government of Nunavut, Wildlife Research Section.

Anderson, M. L., H. D. Cluff, L. D. Mech, and D. R. MacNulty. 2025. "Wolf Density and Predation Patterns in the Canadian High Arctic." *Journal of Wildlife Management*.

Aniśkowicz, B. T., H. Hamilton, D. R. Gray, and C. Downes. 1990. "Nursing Behaviour of Arctic Hares (*Lepus arcticus*)." In *Canada's Missing Dimension, Science and History in the Canadian Arctic Islands*, vol. 2, edited by C. R. Harington, 643–64. Ottawa, ON: Canadian Museum of Nature.

Apps, C. D., and B. N. McLellan. 2006. "Factors Influencing the Dispersion and Fragmentation of Endangered Mountain Caribou Populations." *Biological Conservation* 130:84–97. https://doi.org/10.1016/j.biocon.2005.12.004.

Asa, C. S., L. D. Mech, and U. S. Seal. 1985. "The Use of Urine, Faeces, and Anal-Gland Secretions in Scent-Marking by a Captive Wolf (*Canis lupus*) Pack." *Animal Behaviour* 33:1034–36. https://doi.org/10.1016/S0003-3472(85)80043-9.

Asa, C. S., L. D. Mech, U. S. Seal, and E. D. Plotka. 1990. "The Influence of Social and Endocrine Factors on Urine-Marking by Captive Wolves (*Canis lupus*)." *Hormones and Behavior* 24:497–509. https://doi.org/10.1016/0018-506X(90)90038-Y.

Audlaluk, L. 2021. *What I Remember, What I Know: The Life of a High Arctic Exile*. Toronto: Inhabit Media. https://inhabitbooks.com/products/what-i-remember-what-i-know-the-life-of-a-high-arctic-exile.

Ballard, W. B., L. A. Ayres, C. L. Gardner, and J. W. Foster. 1991. "Den Site Activity Patterns of Gray Wolves, *Canis lupus*, in Southcentral Alaska." *Canadian Field-Naturalist* 105:497–504. https://www.biodiversitylibrary.org/partpdf/358099.

Bass, R. 1992. *The Ninemile Wolves*. New York: Ballantine.

Bekoff, M. 2007 (revised 2024). *The Emotional Lives of Animals*. Novato, CA: New World Library.

Best, T. L, and T. H. Henry. 1994. "*Lepus arcticus*." *Mammal Species* 457:1–9. https://doi.org/10.2307/3504088.

Boitani, L. 2003. "Wolf Conservation and Recovery." In *Wolves: Behavior, Ecology, and Conservation*, edited by L. D. Mech and L. Boitani, 317–40. Chicago: University of Chicago Press.

Bonnyman, S. G. 1975. "Behavioural Ecology of *Lepus arcticus*." MSc thesis, Carleton University, Ottawa. https://doi.org/10.22215/etd/1975-12359.

Bowen, D. 2022. "The People Used as Pawns in Canada's Claim to the Arctic." *Up Here Magazine*, March–April.

Boyd, D. K., D. E. Ausband, H. D. Cluff, J. R. Heffelfinger, J. W. Hinton, B. R. Patterson, and A. P. Wydeven. 2023. "North American Wolves." In *Wild Furbearer Management and Conservation in North America*, edited by T. L. Hiller, R. D. Applegate, R. D. Bluett, S. N. Frey, E. M. Gese, and J. F. Organ, 32.1–32.68. Helena, MT: Wildlife Ecology Institute. https://doi.org/10.59438/FYHC8935.

Breed, M. M., and J. Moore. 2012. *Animal Behavior*. London: Academic Press.

Bump, J. K., T. D. Gable, S. M. Johnson-Bice, A. T. Homkes, D. Freund, S. K. Windels, and S. Chakrabarti. 2022. "Predator Personalities Alter Ecosystem Services." *Frontiers in Ecology and the Environment* 20:275–77. https://doi.org/10.1002/fee.2512.

Bush, D., C. Barry, D. Manson, and N. Burgess. 2015. "Using Grid Cells for Navigation." *Neuron* 87:507–20. https://doi.org/10.1016/j.neuron.2015.07.006.

Carbyn, L. N., S. M. Oosenbrug, and D. W. Anions. 1993. *Wolves, Bison and the Dynamics Related to the Peace Athabasca Delta in Canada's Wood Buffalo National Park*. Canadian Circumpolar Research Series No. 4. Edmonton: Canadian Circumpolar Institute.

Carmichael, L. E., J. Krizan, J. A. Nagy, M. Dumond, D. Johnson, A. Veitch, and C. Strobeck. 2008. "Northwest Passages: Conservation Genetics of Arctic Island Wolves." *Conservation Genetics* 9:879–92.

Caron-Carrier, J., S. Lai, F. Vézina, A. Tam, and D. Berteaux. 2022. "Long-Distance, Synchronized and Directional Fall Movements Suggest Migration in Arctic Hares on Ellesmere Island (Canada)." *Scientific Reports* 12:5003. https://doi.org/10.1038/s41598-022-08347-1.

Cassidy, K. A., D. R. MacNulty, D. R. Stahler, L. D. Mech, and D. W. Smith. 2017. "Sexually Dimorphic Aggression Indicates Male Gray Wolves Specialize in Pack Defense against Conspecific Groups." *Behavioural Processes* 136:64–72. https://doi.org/10.1016/j.beproc.2017.01.011.

Cassidy, K. A., D. R. MacNulty, D. R. Stahler, D. W. Smith, and L. D. Mech. 2015. "Group Composition Effects on Interpack Aggressive Interactions of Gray Wolves in

Yellowstone National Park." *Behavioral Ecology* 26:1352–60. https://doi.org/10.1093/beheco/arv081.

Chapron, G., P. Kaczensky, J. D. C. Linnell, M. von Arx, D. Huber, H. Andren, José Vicente López-Bao,, and Francisco Álvares. 2014. "Recovery of Large Carnivores in Europe's Modern Human-Dominated Landscapes." *Science* 346:1517–19. https://doi.org/10.1126/science.1257553.

Charnov, E. L. 1976. "Optimal Foraging, the Marginal Value Theorem." *Theoretical Population Biology* 9:129–36. https://doi.org/10.1016/0040-5809(76)90040-X.

Ciucci, P., and L. D. Mech. 1992. "Wolf Den Locations in Relation to Winter Territories in Northeastern Minnesota." *Journal of Mammalogy* 73:899–905. https://doi.org/10.2307/1382214.

Clark, K. R. F. 1971. "Food Habits and Behavior of the Tundra Wolf on Central Baffin Island." PhD diss., University of Toronto.

Cluff, H. D., and L. D. Mech. 2023. "A Field Test of GPSeqClus for Establishing Animal Location Clusters." *Ecological Solutions and Evidence* 4:e12204. https://doi.org/10.1002/2688-8319.12204.

COSEWIC (Committee on the Status of Endangered Wildlife in Canada). 2004. *Assessment and Status Report on the Peary Caribou* Rangifer tarandus pearyi *in Canada*. Committee on the Status of Endangered Wildlife in Canada, Ottawa, ON.

———. 2015. *Assessment and Status Report on the Peary Caribou* Rangifer tarandus pearyi *in Canada*. Committee on the Status of Endangered Wildlife in Canada. Ottawa, ON.

Crisler, L. 1958. *Arctic Wild*. New York: Harper and Rowe.

Darwin, C. R. 1859. *The Origin of Species*. London: John Murray.

Dauginis, A. L. A., and L. C. Brown. 2020. "Sea Ice and Snow Phenology in the Canadian Arctic Archipelago from 1997 to 2018." *Arctic Science* 7, no. 1: 192–207. https://doi.org/10.1139/as-2020-0024.

Dawkins, R. 1989. *The Selfish Gene*. Oxford: Oxford University Press.

de Guinea, M., A. Estrada, K. A.-I. Nekaris, and S. Van Belle. 2021. "Cognitive Maps in the Wild: Revealing the Use of Metric Information in Black Howler Monkey Route Navigation." *Journal of Experimental Biology* 224, jeb242430. https://doi.org/10.1242/jeb.242430.

Demma, D. J., and L. D. Mech. 2009. "Wolf Use of Summer Territory in Northeastern Minnesota." *Journal of Wildlife Management* 72:380–84. https://doi.org/10.2193/2008-114.

Dick, L. 2001. *Muskox Land: Ellesmere Island in the Age of Contact*. Calgary: University of Calgary Press.

Eley, T. J. 2014. *Multimedia Atlas of Global Warming and Climatology*. Thousand Oaks, CA: Sage Publications.

Fentress, J. C., and J. Ryon. 1982. "A Long-Term Study of Distributed Pup Feeding in Captive Wolves." In *Wolves of the World: Perspectives of Behavior, Ecology, and Conservation*, edited by F. H. Harrington and P. C. Paquet, 238–61. Park Ridge, NJ: Noyes Publications.

France, R. L. 1993. "The Lake Hazen Trough: A Late Winter Oasis in a Polar Desert." *Biological Conservation* 63:149–51. https://doi.org/10.1016/0006-3207(93)90503-S.

Fredlund, M., J. Boulanger, M. W. Campbell, M. L. Anderson, and C. D. Mallory. 2019. *Distribution and Abundance of Peary Caribou (*Rangifer tarandus pearyi*) and Muskoxen (*Ovibos moschatus*) on Central Ellesmere Island, March 2017*. Igloolik, NU: Nunavut Department of Environment.

Frévol, S. 2019. "Genetic Diversity along the Core-Edge Continuum in a Species with a Continental-Scale Range, *Canis lupus*." MS thesis, Utah State University, Logan.

Frévol, S. A., D. R. MacNulty, M. Anderson., L. E. Carmichael, H. D. Cluff, L. D. Mech, and M. Musiani. 2023. "Geographic Isolation Reduces Genetic Diversity of a Wide-Ranging Terrestrial Vertebrate, *Canis lupus*." *Ecosphere* 14, no. 6: e4536. https://doi.org/10.1002/ecs2.4536.

Fritts, S. H., W. J. Paul, and L. D. Mech. 1985. "Can Relocated Wolves Survive?" *Wildlife Society Bulletin* 13:459–63. https://www.jstor.org/stable/3782671.

Fritts, S. H., R. O. Stephenson, R. D. Hayes, and L. Boitani. 2003. "Wolves and Humans." In *Wolves: Behavior, Ecology, and Conservation*, edited by L. D. Mech and L. Boitani, 289–316. Chicago: University of Chicago Press.

Frost, G. V., U. S. Bhatt, H. E. Epstein, D. A. Walker, M. K. Raynolds, L. T. Berner, J. W. Bjerke, A. L. Breen, B. C. Forbes, S. J. Goetz, et al. 2019. "Tundra Greenness." In *Arctic Report Card 2019*, edited by J. Richter-Menge, M. L. Druckenmiller, and M. Jeffries, 48–57. Washington, DC: National Oceanic and Atmospheric Administration. https://doi.org/10.25923/8n78-wp73.

Fuller, T. K., L. D. Mech, and J. Fitts-Cochran. 2003. "Population Dynamics." In *Wolves: Behavior, Ecology, and Conservation*, edited by L. D. Mech and L. Boitani, 161–91. Chicago: University of Chicago Press.

Fuller, W. A., and N. S. Novakowski. 1955. "Wolf Control Operations, Wood Buffalo National Park, 1951–1952." *Wildlife Management Bulletin* Series 1, No. 11. Ottawa: Canadian Wildlife Service.

Garland, T. 1983. "The Relation between Maximal Running Speed and Body Mass in Terrestrial Mammals." *Journal of Zoology* 199:1557–70. https://doi.org/10.1111/j.1469-7998.1983.tb02087.x.

Gipson, P. S., W. B. Ballard, R. M. Nowak, and L. D. Mech. 2000. "Accuracy and Precision of Estimating Age of Gray Wolves by Tooth Wear." *Journal of Wildlife Management* 64:752–58. https://www.jstor.org/stable/3802745.

Gitay, H., A. Suarez, R. T. Watson, and D. J. Dokken. 2002. *Climate Change and Biodiversity*, Intergovernmental Panel on Climate Change, Technical Paper V, 2, 10–11. World Meterological Organization and United Nations Environment Programme. https://archive.ipcc.ch/pdf/technical-papers/climate-changes-biodiversity-en.pdf.

Gloveli, N., J. Simonnet, W. Tang, M. Concha-Miranda, E. Maier, A. Dvorzhak, D. Schmitz, and M. Brecht. 2023. "Play and Tickling Responses Map to the Lateral Columns of the Rat Periaqueductal Gray." *Neuron* 111, no. 19, P3041-3052.e7. https://doi.org/10.1016/j.neuron.2023.06.018.

Goldman, J. G. 2011. "Rats, Bees, and Brains: The Death of the 'Cognitive Map.'" *Scientific American*, July 12. https://www.scientificamerican.com/blog/thoughtful-animal/httpblogsscientificamericancomthoughtful-animal20110712rats-bees-and-brains-the-death-of-the-cognitive-map/.

Goodman, P. A., and E. Klinghammer. 1985. *Wolf Ethogram*. Ethology Series 3. Battle Ground, IN: North American Wildlife Park Foundation.

Grace, E. S. 1976. "Interactions between Men and Wolves at an Arctic Outpost on Ellesmere Island." *Canadian Field-Naturalist* 90:149–56.

Gray, D. R. 1970. "The Killing of a Bull Muskox by a Single Wolf." *Arctic* 23:197–99. https://www.jstor.org/stable/40507733.

———.1983. "Interactions between Wolves and Muskoxen on Bathurst Island, Northwest Territories, Canada." *Annales Zoologici Fennici* 174:255–57.

———. 1987. *The Muskoxen of Polar Bear Pass*. Markham, Ontario: Fitzhenry and Whiteside.

———. 1993. "Behavioural Adaptations to Arctic Winter: Shelter Seeking by Arctic Hare (*Lepus arcticus*)." *Arctic* 46:340–53. https://www.jstor.org/stable/40511436.

Griffin, K. R., G. H. Roffler, and E. M. Dymit. 2023. "Wolves on the Katmai Coast Hunt Sea Otters and Harbor Seals." *Ecology* 104, no. 23: e4185. https://doi.org/10.1002/ecy.4185.

Haber, G. C. 1977. "Socio-Ecological Dynamics of Wolves and Prey in a Subarctic Ecosystem." PhD diss., University of British Columbia, Vancouver. https://dx.doi.org/10.14288/1.0094168.

Hamilton. W. D. 1964. "The Genetical Evolution of Social Behaviour." *Journal of Theoretical Biology* 7:1–16. https://doi.org/10.1016/0022-5193(64)90039-6.

———. 1971. "Geometry for a Selfish Herd." *Journal of Theoretical Biology* 31:295–311. https://doi.org/10.1016/0022-5193(71)90189-5.

Harrington, F. H. 1981a. "Urine-Marking and Caching Behavior in the Wolf." *Behaviour* 76:280–88.

———. 1981b. "Urine-Marking and Its Relation to Caching in Coyotes." Paper presented at the Animal Behavior Society Meeting, Woods Hole, MA.

Harrington, F. H., and C. S. Asa. 2003. "Wolf Communication." In *Wolves: Behavior, Ecology, and Conservation*, edited by L. D. Mech and L. Boitani, 66–103. Chicago: University of Chicago Press.

Harrington, F. H., and L. D. Mech. 1979. "Wolf Howling and Its Role in Territory Maintenance." *Behaviour* 68:207–49. https://doi.org/10.1163/156853979X00322.

Harrington, F. H., P. C. Paquet, J. Ryon, and J. C. Fentress. 1982. "Monogamy in Wolves: A Review of the Evidence." In *Wolves of the World: Perspectives on Behaviour, Ecology and Conservation*, edited by F. H. Harrington and P. C. Paquet, 209–22. Park Ridge, NJ: Noyes Publications.

Harten, L., A. Katz, A. Goldshtein, M. Handel, and Y. Yovel. 2020. "The Ontogeny of a Mammalian Cognitive Map in the Real World." *Science* 369:194–97. https://www.science.org/doi/10.1126/science.aay3354.

Hayes, R. D., R. Farnell, R. M. P. Ward, J. Carey, M. Dehn, G. W. Kuzyk, A. M. Baer, C. L. Gardner, and M. O'Donoghue. 2003. "Experimental Reduction of Wolves in the Yukon: Ungulate Responses and Management Implications." *Wildlife Monographs* 152:1–35. https://www.jstor.org/stable/3830836.

Heard, D. C. 1992. "The Effect of Wolf Predation and Snow Cover on Musk-Ox Group Size." *American Naturalist* 139:190–204. https://doi.org/10.1086/285320.

Hearn, B. J., L. B. Keith, and O. J. Rongstad. 1987. "Demography and Ecology of the Arctic Hare (*Lepus arcticus*) in Southwestern Newfoundland." *Canadian Journal of Zoology* 65:852–61. https://doi.org/10.1139/z87-136.

Hendricks, S. A., R. M. Schweizer, and R. K. Wayne. 2019. "Conservation Genomics Illuminates the Adaptive Uniqueness of North American Gray Wolves." *Conservation Genetics* 20:29–43. https://doi.org/10.1007/s10592-018-1118-z.

Henry, J. D. 1977. "The Use of Urine Marking in the Scavenging Behavior of the Red Fox (*Vulpes vulpes*)." *Behaviour* 61:82–106. https://doi.org/10.1163/156853977X00496.

Hervieux, D., M. Hebblewhite, D. Stepnisky, M. Bacon, and S. Boutin. 2014. "Managing Wolves (*Canis lupus*) to Recover Threatened Woodland Caribou (*Rangifer tarandus caribou*) in Alberta." *Canadian Journal of Zoology* 92:1029–1037. https://doi.org/10.1139/cjz-2014-0142.

Holt, R. D. 1977. "Predation, Apparent Competition, and the Structure of Prey Communities." *Theoretical Population Biology* 12:197–229. https://doi.org/10.1016/0040-5809(77)90042-9.

Hone, E. 1934. "The Present Status of the Muskox in Arctic North America and Greenland: With Notes on Distribution, Extirpation, Transplantation, Protection, Habits and Life History." *American Committee for International Conservation*. Special Publication No. 5.

Howell, S. E. L., F. Laliberté, R. Kwok, C. Derksen, and J. King. 2016. "Landfast Ice Thickness in the Canadian Arctic Archipelago from Observations and Models." *Cryosphere* 10:1463–75. https://doi.org/10.5194/tc-10-1463-2016.

Hubert, B. 1974. "Estimated Productivity of Muskox *Ovibos moschatus* Northeastern Island, NWT." MS thesis, University of Manitoba, Winnipeg.

Jędrzejewski, W., K. Schmidt, J. Theuerkauf, B. Jędrzejewska, and H. Okarma. 2001. "Daily Movements and Territory Use by Radio-Collared Wolves, *Canis lupus*, in Bialowieza Primeval Forest in Poland." *Canadian Journal of Zoology* 79:1–12. https://doi.org/10.1139/z01-147.

Jenkins, D., M. Campbell, G. Hope, J. Goorts, and P. McLoughlin. 2011. *Recent Trends in Abundance of Peary Caribou (*Rangifer tarandus pearyi*) and Muskoxen (*Ovibos moschatus*) in the Canadian Arctic Archipelago, Nunavut*. Pond Inlet, NU: Nunavut Department of Environment. https://www.gov.nu.ca/sites/default/files/publications/2022-01/recent_trends_in_abundance_of_peary_caribou_and_muskoxen_in_the_canadian_arctic_archipelago_nunavut_final_report_june_10_2011.pdf.

Joly K., M. S. Sorum, and M. D. Cameron. 2018. "Denning Ecology of Wolves in East-Central Alaska, 1993–2017." *Arctic* 71:444–55. https://www.jstor.org/stable/26567073.

Joslin, P. W. B. 1966. "Summer Activities of Two Timber Wolf (*Canis lupus*) Packs in Algonquin Park." MS thesis, University of Toronto.

Karl, T. 1999. "Weather and Climate Extremes: Changes, Variations and a Perspective from the Insurance Industry—Overview." *Climate Change* 42:1–2.

Karl, T. R., N. Nicholls, and J. Gregory. 1997. "The Coming Climate." *Scientific American*, May, 78–83. https://www.jstor.org/stable/24993744.

Kelsall, J. P. 1968. *The Migratory Barren-Ground Caribou of Canada*. Canadian Wildlife Service. Ottawa, ON: Queen's Printer.

Kim, Y. H., S. K. Min, N. P. Gillett, Dirk Notz, and Elizaveta Malinina. 2023. "Observationally-Constrained Projections of an Ice-Free Arctic Even under a Low Emission Scenario." *Nature Communications* 14:3139. https://doi.org/10.1038/s41467-023-38511-8.

Kleiman, D. 1966. "Scent Marking in the Canidae." *Symposia of the Zoological Society of London* 18:167–77.

Klinghammer, E., and P. A. Goodmann. 1987. "Socialization and Management of Wolves in Captivity." In *Man and Wolf*, edited by H. Frank, 31–59. Dordrecht: Dr. W. Junk Publishers.

Kreeger, T. J. 2003. "The Internal Wolf: Physiology, Pathology and Pharmacology." In *Wolves: Behavior, Ecology, and Conservation*, edited by L. D. Mech and L. Boitani, 192–217. Chicago: University of Chicago Press.

Lai, S., É. Desjardins, J. Caron-Carrier, C. Couchoux, F. Vézina, A. Tam, N. Koutroulides, and D. Berteaux. 2022. "Unsuspected Mobility of Arctic Hares Revealed by Longest Journey Ever Recorded in a Lagomorph." *Ecology* 103, no. 3: e3620. https://doi.org/10.1002/ecy.3620.

Laikre, L., and N. Ryman. 1991. "Inbreeding Depression in a Captive Wolf (*Canis lupus*) Population." *Biological Conservation* 5:33-40. https://doi.org/10.1111/j.1523-1739.1991.tb00385.x.

Latham, A. D. M., and S. Boutin. 2011. "Wolf, *Canis lupus*, Pup Mortality: Interspecific Predation or Non-Parental Infanticide?" *Canadian Field-Naturalist* 125:158–61. https://doi.org/10.22621/cfn.v125i2.1199.

Latour, P. B. 1987. "Observations on Demography, Reproduction, and Morphology of Muskoxen (*Ovibos moschatus*) on Banks Island, Northwest Territories." *Canadian Journal of Zoology* 65:265–69. https://doi.org/10.1139/z87-041.

Lent, P. 1978. "Muskox." In *Big Game of North America: Ecology and Management*, edited by J. L. Schmid and D. L. Gilbert, 135–47. Harrisburg, PA: Wildlife Management Institute and Stackpole Books.

Lesins, G., T. J. Duck, and J. R. Drummond. 2010. "Climate Trends at Eureka in the Canadian High Arctic." *Atmosphere-Ocean* 48:59–80. https://doi.org/10.3137/AO1103.2010.

Leutgeb, S., J. K. Leutgeb, C. A. Barnes, E. I. Moser, B. L. McNaughton, and M.-B. Moser. 2005. "Independent Codes for Spatial and Episodic Memory in Hippocampal Neuronal Ensembles." *Science* 309:619–23. https://doi.org/10.1126/science.1114037.

Linnell, J. D. C., E. Kovtun, and I. Rouart. 2021. *Wolf Attacks on Humans: An Update for 2002–2020.* NINA Report 1944. Trondheim, Norway: Norwegian Institute for Nature Research (NINA). https://hdl.handle.net/11250/2729772.

Lonardo, L., C. J. Völter, C. Lamm, and L. Huber. 2021. "Dogs Follow Human Misleading Suggestions More Often When the Informant Has a False Belief." *Proceedings of the Royal Society B: Biological Sciences* 288:20210906. https://doi.org/10.1098/rspb.2021.0906.

MacNulty, D. R. 2002. "The Predatory Sequence and the Influence of Injury Risk on Hunting Behavior in the Wolf." MS thesis, University of Minnesota, St. Paul.

MacNulty, D. R., L. D. Mech, and D. W. Smith. 2007. "A Proposed Ethogram of Large-Carnivore Predatory Behavior, Exemplified by the Wolf." *Journal of Mammalogy* 88:595–605. https://doi.org/10.1644/06-MAMM-A-119R1.1.

MacNulty, D. R., D. W. Smith, J. A. Vucetich, L. D. Mech, D. R. Stahler, and C. Packer. 2009. "Predatory Senescence in Aging Wolves." *Ecology Lettters* 12:1–10. https://doi.org/10.1111/j.1461-0248.2009.01385.x.

Magoun, A. 1976. "Summer Scavenging Activity in Northeastern Alaska." MS thesis, University of Alaska, Fairbanks. http://www.adfg.alaska.gov/static/home/library/pdfs/wildlife/research_pdfs/summer_scavenging_activity_northeastern_alaska.pdf.

Mallory, C. D., M. Fredlund, and M. W. Campbell. 2020. "Apparent Collapse of the Peary Caribou (*Rangifer tarandus pearyi*) Population on Axel Heiberg Island, Nunavut, Canada." *Arctic* 73:499–508. https://www.jstor.org/stable/26991437.

Marquard-Petersen, U. 2008. "Reproduction and Mortality of the High Arctic Wolf, *Canis lupus arctos*, in Northeast Greenland, 1978–1998." *Canadian Field-Naturalist* 122:142–52. https://doi.org/10.22621/cfn.v122i2.573.

———. 2022. Behaviors of High Arctic Wolves in Response to Humans. *Arctic* 75:378–89. https://doi.org/10.14430/arctic75966.

Martin, H., L. D. Mech, J. Fieberg, M. Metz, D. R. MacNulty, D. R. Stahler, and D. W. Smith 2018. "Factors Affecting Gray Wolf (*Canis lupus*) Encounter Rate with Elk (*Cervus elaphus*) in Yellowstone National Park." *Canadian Journal of Zoology* 96:1032–42. https://doi.org/10.1139/cjz-2017-0220.

McCullough, D. R., and D. E. Ullrey. 1983. "Proximate Mineral and Gross Energy Composition of White-Tailed Deer." *Journal of Wildlife Management* 473:430–41. https://doi.org/10.2307/3808516.

McIntyre, R. 2013. "The 06 Female." *In Wild Wolves We Have Known*, edited by R. P. Thiel, A. C. Thiel, and M. Strozewski, 123–30. Minneapolis, MN: International Wolf Center.

———. 2019. *The Rise of Wolf 8: Witnessing the Triumph of Yellowstone's Underdog.* Vancouver, BC: Greystone Books.

———. 2020. *The Reign of Wolf 21: The Saga of Yellowstone's Legendary Druid Pack.* Vancouver, BC: Greystone Books.

———. 2021. *The Redemption of Wolf 302: From Renegade to Yellowstone Alpha Male.* Vancouver, BC: Greystone Books.

———. 2022. *The Alpha Female Wolf: The Fierce Legacy of Yellowstone's 06.* Vancouver, BC: Greystone Books.

McIntyre, R., and L. D. Mech. 2024. "Multiple Same-Sex Scent-Marking in Free Ranging Gray Wolf Packs." *Northwestern Naturalist* 105, no. 1.

McIntyre, R., J. B. Theberge, M. T. Theberge, and D. W. Smith. 2017. "Behavioral and Ecological Implications of Seasonal Variation in the Frequency of Daytime Howling by Yellowstone Wolves." *Journal of Mammalogy* 98:827–34. https://doi.org/10.1093/jmammal/gyx034.

McLeod. P. J. 1990. "Infanticide by Female Wolves." *Canadian Journal of Zoology* 68:402–404. https://doi.org/10.1139/z90-058.

Mech, L. D. 1966. *The Wolves of Isle Royale*. National Parks Fauna Series No. 7. Washington, D.C.: US Government Printing Office.

———. 1970. *The Wolf: The Ecology and Behavior of an Endangered Species*. New York: Natural History Press and Doubleday Publishing.

———. 1987. "At Home with the Arctic Wolf." *National Geographic* 171:562–93. https://drive.google.com/file/d/0B9bX852JMJ__YTQwNTU1OWYtNjM1Zi00NzJiLWJlNWQtOWE3NTIwY2ZhOGEw/view?resourcekey=0-CfqzjUrJodyJDr4v56jUrg.

———. 1988a. *The Arctic Wolf: Living with the Pack*. Stillwater, MN: Voyageur Press.

———. 1988b. "Life in the High Arctic." *National Geographic* 173:759–67.

———. 1988c. "Longevity in Wild Wolves." *Journal of Mammalogy* 69:197–98. https://doi.org/10.2307/1381776.

———. 1993a. "Details of a Confrontation between Two Wild Wolves." *Canadian Journal of Zoology* 71:1900–1903. https://doi.org/10.1139/z93-271.

———. 1993b. "Resistance of Young Wolf Pups to Inclement Weather." *Journal of Mammalogy* 74:485–86. https://doi.org/10.2307/1382407.

———. 1994a. "Buffer Zones of Territories of Gray Wolves as Regions of Intraspecific Strife." *Journal of Mammalogy* 75:199–202. https://doi.org/10.2307/1382251.

———. 1994b. "Regular and Homeward Travel Speeds of Arctic Wolves." *Journal of Mammalogy* 75:741–42. https://doi.org/10.2307/1382524.

———. 1995a. "Summer Movements and Behavior of an Arctic Wolf, *Canis lupus*, Pack without Pups." *Canadian Field-Naturalist* 109:473–75.

———. 1995b. "A Ten-Year History of the Demography and Productivity of an Arctic Wolf Pack." *Arctic* 48:329–32. https://www.jstor.org/stable/40511934.

———. 1997a. *The Arctic Wolf: Ten Years with the Pack*. Stillwater, MN: Voyageur Press.

———. 1997b. "An Example of Endurance in an Old Wolf, *Canis lupus*." *Canadian Field-Naturalist* 111:654–55.

———. 1999. "Alpha Status, Dominance, and Division of Labor in Wolf Packs." *Canadian Journal of Zoology* 77:1196–203. https://doi.org/10.1139/z99-099.

———. 2000a. "Lack of Reproduction in Musk Oxen and Arctic Hares Caused by Early Winter?" *Arctic* 53:69–71.

———. 2000b. "Leadership in Wolf, *Canis lupus*, Packs." *Canadian Field-Naturalist* 114:259–63. http://digitalcommons.unl.edu/usgsnpwrc/384.

———. 2001. "Standing Over and Hugging in Wild Wolves." *Canadian Field-Naturalist* 115:179–81. https://doi.org/10.5962/p.363766.

———. 2002. "Breeding Season of Wolves, *Canis lupus*, in Relation to Latitude." *Canadian Field-Naturalist* 116:139–40. https://doi.org/10.5962/p.363419.

———. 2004. "Is Climate Change Affecting Wolf Populations in the High Arctic?" *Climate Change* 67:87–93. https://doi.org/10.1007/s10584-004-7093-z.

———. 2005. "Decline and Recovery of a High Arctic Wolf-Prey System." *Arctic* 58:305–7. https://www.jstor.org/stable/40512716.

———. 2006. "Urine-Marking and Ground-Scratching by Free-Ranging Arctic Wolves, *Canis lupus arctos*, in Summer." *Canadian Field-Naturalist* 120:466–70. https://doi.org/10.22621/cfn.v120i4.356.

———. 2007a. "Annual Arctic Wolf Pack Size Related to Arctic Hare Numbers." *Arctic* 60:309–11. https://www.jstor.org/stable/40512898.

———. 2007b. "Possible Use of Foresight, Understanding, and Planning by Wolves Hunting Muskoxen." *Arctic* 60:145–49. https://www.jstor.org/stable/40513130.

———. 2010. "Proportion of Calves and Adult Muskoxen, *Ovibos moschatus*, Killed by Gray Wolves (*Canis lupus*) in July on Ellesmere Island." *Canadian Field-Naturalist* 124:258–60. https://doi.org/10.22621/cfn.v124i3.1083.

———. 2011. "Gray Wolf (*Canis lupus*) Movements around a Kill Site and Implications for GPS Collar Studies." *Canadian Field-Naturalist* 125:353–56. https://doi.org/10.22621/cfn.v125i4.1263.

———. 2013. "Brutus." In *Wild Wolves We Have Known*, edited by R. P. Theil, A. C. Theil, and M. Strozewski, 216–25. Minneapolis, MN: International Wolf Center.

———. 2014. "A Gray Wolf (*Canis lupus*) Delivers Live Prey to a Pup." *Canadian Field-Naturalist* 128:189–90. https://doi.org/10.22621/cfn.v128i2.1584.

———. 2017a. "Extinguishing a Learned Response in a Wild Wolf." *Canadian Field-Naturalist* 131:23–25. http://dx.doi.org/10.22621/cfn.v131i1.1951.

———. 2017b. "Where Can Wolves Live and How Can We Live with Them?" *Biological Conservation* 210:310–17. https://doi.org/10.1016/j.biocon.2017.04.029.

———. 2022. "Newly Documented Behavior of Free-Ranging Arctic Wolf Pups." *Arctic* 75:272–76. https://doi.org/10.14430/arctic75056.

———. 2024. "Plural Breeding in Wolf (*Canis lupus*) Packs: How Often?" *Canadian Field Naturalist.*

Mech, L. D., and L. G. Adams. 1999. "Killing of a Muskox, *Ovibos moschatus*, by Two Wolves, *Canis lupus*, and Subsequent Caching." *Canadian Field-Naturalist* 113:673–75. https://doi.org/10.5962/p.358674.

Mech, L. D., L. G. Adams. T. J. Meier, J. W. Burch, and B. W. Dale. 1998. *The Wolves of Denali.* Minneapolis: University of Minnesota Press.

Mech, L. D., and S. M. Barber-Meyer. 2015. "Yellowstone Wolf (*Canis lupus*) Density Predicted by Elk (*Cervus elaphus*) Biomass." *Canadian Journal of Zoology* 93:499–502. https://doi.org/10.1139/cjz-2015-0002.

———. 2017. "Seasonality of Intraspecific Mortality by Gray Wolves." *Journal of Mammalogy* 98:1538–46. https://doi.org/10.1093/jmammal/gyx113.

Mech, L. D., S. M. Barber-Meyer, and J. Erb. 2016. "Wolf (*Canis lupus*) Generation Time and Proportion of Current Breeding Females by Age." *PLoS One* 11:e0156682. https://doi.org/10.1371/journal.pone.0156682.

Mech, L. D., and L. Boitani. 2003. "Wolf Social Ecology." In *Wolves: Behavior, Ecology, and Conservation*, edited by L. D. Mech and L. Boitani, 1–34. Chicago: University of Chicago Press.

Mech, L. D., and G. Breining. 2020. *Wolf Island*. Minneapolis: University of Minnesota Press.

Mech, L. D., and H. D. Cluff. 2009. "Long Daily Movements of Wolves (*Canis lupus*) during Pup Rearing." *Canadian Field-Naturalist* 123:68–69. https://doi.org/10.22621/cfn.v123i1.675.

———. 2010. "Prolonged Intensive Dominance Behavior between Gray Wolves, *Canis lupus*." *Canadian Field-Naturalist* 124:215–18. https://doi.org/10.22621/cfn.v124i3.1076.

———. 2011. "Movements of Wolves at the Northern Extreme of the Species' Range Including during Four Months of Darkness." *PLoS One* 6:e25328. https://doi.org/10.1371/journal.pone.0025328.

Mech, L. D., and G. D. DelGiudice. 1985. "Limitations of the Marrow-Fat Technique as an Indicator of Condition." *Wildlife Society Bulletin* 13:204–6. https://www.jstor.org/stable/3781442.

Mech, L. D., L. D. Frenzel Jr., and P. D. Karns. 1971. "The Effect of Snow Conditions on the Ability of Wolves to Capture Deer." In *Ecological Studies of the Timber Wolf in Northeastern Minnesota*, edited by L. D. Mech and L. D. Frenzel Jr., 51–59. USDA Forest Service Research Paper NC-52. St. Paul, MN: North Central Forest Experimental Station. https://www.fs.usda.gov/research/treesearch/10573.

Mech, L. D., and H. Hertel. 1983. "An Eight-Year Demography of a Minnesota Wolf Pack." *Acta Zoologica Fennica* 174:249–50.

Mech, L. D., and L. Janssens. 2022. "An Assessment of Current Wolf Domestication Hypotheses Based on Wolf Ecology and Behaviour." *Mammal Review* 52:304–14. https://doi.org/10.1111/mam.12273.

Mech, L. D., and R. McIntyre. 2022. "Key Observations of Flexed-Leg Urinations in Free-Ranging Gray Wolf (*Canis lupus*)." *Canadian Field-Naturalist* 136:10–12. https://doi.org/10.22621/cfn.v136i1.2781.

———. 2023. "An Observation of Incest Avoidance in Gray Wolves (*Canis lupus*)." *Canadian-Field Naturalist* 137, no. 3. https://doi.org/10.22621/cfn.v137i3.2971.

Mech, L. D., and S. B. Merrill. 1998. "Daily Departure and Return Patterns of Wolves, *Canis lupus*, from a Den at 80° N Latitude." *Canadian Field-Naturalist* 112:515–17.

Mech, L. D., and J. M. Packard. 1990. "Possible Use of Wolf (*Canis lupus*) Den over Several Centuries." *Canadian Field-Naturalist* 104:484–85. http://dx.doi.org/10.5962/p.356419.

Mech, L. D., and R. O. Peterson. 2003. "Wolf-Prey Relations." In *Wolves: Behavior, Ecology, and Conservation*, edited by L. D. Mech and L. Boitani, 131–57. Chicago: University of Chicago Press.

Mech, L. D., M. K. Phillips, D. W. Smith, and T. J. Kreeger. 1996. "Denning Behaviour of Non-Gravid Wolves, *Canis lupus*." *Canadian Field-Naturalist* 110:343–45.

Mech, L. D., U. S. Seal, and S. M. Arthur. 1984. "Recuperation of a Severely Debilitated Wolf." *Journal of Wildlife Diseases* 20:166–68. https://doi.org/10.7589/0090-3558-20.2.166.

Mech, L. D., D. W. Smith, and D. R. MacNulty. 2015. *Wolves on the Hunt: The Behavior of Wolves Hunting Wild Prey*. Chicago: University of Chicago Press.

Mech, L. D., P. C. Wolf, and J. M. Packard. 1999. "Regurgitative Food Transfer among Wild Wolves." *Canadian Journal of Zoology* 77:1192–195. https://doi.org/10.1139/z99-097.

Merrill, S. B., L. G. Adams, M. E. Nelson, and L. D. Mech. 1998. "Testing Releasable GPS Collars on Wolves and White-Tailed Deer." *Wildlife Society Bulletin* 26:830–35. https://www.jstor.org/stable/3783557.

Mertl-Millhollen, A. S., P. A. Goodmann, and E. Klinghammer. 1986. "Wolf Scent Marking with Raised-Leg Urination." *Zoo Biology* 5:7–20. https://doi.org/10.1002/zoo.1430050103.

Miller, F. 1978. "Interactions between Men, Dogs, and Wolves on Western Queen Elizabeth Island, Northwest Territories, Canada." *Musk-ox* 22:70–72.

———. 1995. "Status of Wolves in the Canadian Arctic Islands." In *Ecology and Conservation of Wolves in a Changing World*, edited by L. N. Carbyn, S. H. Fritts, and D. R. Seip, 35–42. Occasional Publication No. 35. Edmonton, AB: Canadian Circumpolar Institute.

Miller, F. L., and S. J. Barry. 2009. "Long-Term Control of Peary Caribou Numbers by Unpredictable, Exceptionally Severe Snow or Ice Conditions in a Non-Equilibrium Grazing System." *Arctic* 62:175–89. https://www.jstor.org/stable/40513286.

Miller, F. L., and A. Gunn. 2003. "Catastrophic Die-Off of Peary Caribou on the Western Queen Elizabeth Islands, Canadian High Arctic." *Arctic* 56:381–90. https://www.jstor.org/stable/40513077.

Miller, F. L., and F. D. Reintjes. 1995. "Wolf-Sightings on the Canadian Arctic Islands." *Arctic* 48:313–23. https://www.jstor.org/stable/40511932.

Miller K., K. Tietjen, and K. C. Beard. 2023. "Basal Primatomorpha Colonized Ellesmere Island (Arctic Canada) during the Hyperthermal Conditions of the Early Eocene Climatic Optimum." *PLoS ONE* 18, no.1: e0280114. https://doi.org/10.1371/journal.pone.0280114.

Munthe, K., and J. H. Hutchison. 1978. "A Wolf-Human Encounter on Ellesmere Island, Canada." *Journal of Mammalogy* 59:876–78. https://doi.org/10.2307/1380164.

Murie, A. 1944. *The Wolves of Mount McKinley*. US National Park Service Fauna Series, No. 5. Washington, DC: US Government Printing Office. https://www.jstor.org/stable/j.ctvcwnw0d.

Muro, C., R. Escobedo, L. Spector, and R. P. Coppinger. 2011. "Wolf-Pack (*Canis lupus*) Hunting Strategies Emerge from Simple Rules in Computational Simulations." *Behavioural Processes* 88:192–97. https://doi.org/10.1016/j.beproc.2011.09.006.

Nelson, M. E., and L. D. Mech. 1981. "Deer Social Organization and Wolf Depredation in Northeastern Minnesota." *Wildlife Monographs* No. 77. https://www.jstor.org/stable/3830578.

———. 1993. "Prey Escaping Wolves Despite Close Proximity." *Canadian Field-Naturalist* 107:245–46.

Nunez, I. B., and F. J. de Miguel. 2004. "Variation in Stimulus, Seasonal Context, and Response to Urine Marks by Captive Iberian Wolves (*Canis lupus signatus*)." *Acta Ethologica* 7:51–57. https://doi.org/10.1007/s10211-004-0101-5.

Ortiz-Leal, I., M. V. Torres, J.-D. Barreiro-Vázquez, A. López-Beceiro, L. Fidalgo, and P. Sanchez-Quinteir. 2024. "Sensory Adaptations: Insights into the Vomeronasal System of the Iberian Wolf." *Journal of Anatomy*. https://doi.org/10.1111/joa.14024.

Packard, J. M. 2003. "Wolf Behavior: Reproductive, Social, and Intelligent." In *Wolves: Behavior, Ecology, and Conservation*, edited by L. D. Mech and L. Boitani, 35–65. Chicago: University of Chicago Press.

Packard, J. M., L. D. Mech, and R. R. Ream. 1992. "Weaning in an Arctic Wolf Pack: Behavioural Mechanisms." *Canadian Journal of Zoology* 70:1269–1275. https://doi.org/10.1139/z92-177.

Palacios, V., E. Font, and R. Marquez. 2007. "Iberian Wolf Howls: Acoustic Structure, Individual Variation, and a Comparison with North American Populations." *Journal of Mammalogy* 88:606–13. https://doi.org/10.1644/06-MAMM-A-151R1.1.

Palacios, V., and L. D. Mech. 2010. "Problems with Studying Wolf Predation on Small Prey in Summer via Global Positioning Collars." *European Journal of Wildlife Research* 57:149–56. https://doi.org/10.1007/s10344-010-0408-7.

Pappas, S. 2023. "Is the Alpha Wolf Idea a Myth?" *Scientific American*, Feb. 28. https://www.scientificamerican.com/article/is-the-alpha-wolf-idea-a-myth/.

Paquet, P. C. 1991. "Scent-Marking Behavior of Sympatric Wolves (*Canis lupus*) and Coyotes (*C. latrans*) in Riding Mountain National Park." *Canadian Journal of Zoology* 69:1721–27. https://doi.org/10.1139/z91-240.

Parker, G. R. 1977. "Morphology, Reproduction, Diet, and Behavior of the Arctic Hare (*Lepus arcticus monstrabilis*) on Axel Heiberg Island, Northwest Territories." *Canadian Field-Naturalist* 91:8–18. https://doi.org/10.5962/p.345322.

Parmelee, D. F. 1964. "Myth of the Wolf." *Beaver* 295:4–9.

Pasitschniak-Arts, M., M. E. Taylor, and L. D. Mech. 1988. "Note on Skeletal Injuries in an Adult Arctic Wolf." *Arctic Alpine Research* 20:360–65. https://doi.org/10.2307/1551269.

Peters, R. 1979. "Mental Maps in Wolf Territoriality." In *The Behavior and Ecology of Wolves*, edited by E. Klinghammer, 119–52. New York: Garland STPM Press.

Peters, R., and L. D. Mech. 1975. "Scent-Marking in Wolves: A Field Study." *American Scientist* 63:628-637. https://www.jstor.org/stable/27845779.

Peterson, R. O. 1977. *Wolf Ecology and Prey Relationships on Isle Royale*. National Park Service Monograph Series 11. Washington, DC: United States Department of the Interior. https://www.nps.gov/parkhistory/online_books/science/11/contents.htm.

Peterson, R. O., and D. L. Allen. 1974. "Snow Conditions as a Parameter in Moose-Wolf Relationships." *Naturaliste Canadien* (Quebec) 101:481–84.

Peterson, R. O., and P. Ciucci. 2003. "The Wolf as a Carnivore." In *Wolves: Behavior, Ecology, and Conservation*, edited by L. D. Mech and L. Boitani, 104–30. Chicago: University of Chicago Press.

Peterson, R. O., A. Jacobs, T. D. Drummer, L. D. Mech, and D. W. Smith. 2002. "Leadership Behavior in Relation to Dominance and Reproductive Status in Gray Wolves." *Canadian Journal of Zoology* 80:1405–1412. https://doi.org/10.1139/z02-124.

Plint, T., F. J. Longstaffe, A. Ballantyne, A. Telka, and N. Rybczynski. 2020. "Evolution of Woodcutting Behaviour in Early Pliocene Beaver Driven by Consumption of Woody Plants." *Scientific Reports* 10:13111. https://doi.org/10.1038/s41598-020-70164-1.

Popper, I. 2017. "A Bone to Pick: One Pack's Drama over Feeding an Old Wolf." *International Wolf* 27:24–27.

Post, E., E. Kaarlejarvi, M. Macias-Fauria, D. A. Watts, P. S. Bøving, S. M. P. Cahoon, R. Conor Higgins, C. John, J. T. Kerby, C. Pedersen, et al. 2023. "Large Herbivore Diversity Slows Sea-Ice Associated Decline in Arctic Tundra Diversity." *Science* 380:1282–87. https://www.science.org/doi/10.1126/science.add2679.

Powell, R. A., S. A. Mansfield, and L. L. Rogers. 2023. "Comparison of Behaviors of Black Bears with and without Habituation to Humans and Supplemental Research Feeding." *Journal of Mammalogy* 103:1350–63. https://doi.org/10.1093/jmammal/gyac081.

Pulliainen, E. 1965. "Studies on the Wolf (*Canis lupus* L.) in Finland." *Annales Zoologici Fennici* 2:215–59. https://www.jstor.org/stable/23730710.

Qikiqtani Inuit Association. 2014. *Qikiqtani Truth Commission: Community Histories 1950–1975*. Iqaluit, Nunavut: Inhabit Media. https://www.qtcommission.ca/en/reports/communities.

Rantanen, M., A. Y. Karpechko, A. Lipponen, Kalle Nordling, Otto Hyvärinen, Kimmo Ruosteenoja, Timo Vihma, and Ari Laaksonen. 2022. "The Arctic Has Warmed Nearly Four Times Faster Than the Globe since 1979." *Communications Earth & Environment* 3:168. https://doi.org/10.1038/s43247-022-00498-3.

Reicher, V., A. Bálint, D. Újváry, and M. Gácsi. 2022. "Non-Invasive Sleep EEG Measurement in Hand Raised Wolves." *Scientific Reports* 12:9792. https://www.nature.com/articles/s41598-022-13643-x.

Reynolds, P. E. 1993. "Dynamics of Muskox Groups in Northeastern Alaska." *Rangifer* 13:83–89. https://doi.org/10.7557/2.13.2.1082.

Rothman, R. J., and L. D. Mech. 1979. "Scent-Marking in Lone Wolves in and Newly Formed Pairs." *Animal Behaviour* 27:750–60. https://doi.org/10.1016/0003-3472(79)90010-1.

Ruprecht, J. S., Ausband, D. E., Mitchell, M. S., Garton, E. O., and Zager, P. 2012. "Homesite Attendance Based on Sex, Breeding Status, and Number of Helpers in Gray Wolf Packs." *Journal of Mammalogy* 93:1001–5. https://doi.org/10.1644/11-MAMM-A-330.1.

Ryon, J., and R. E. Brown. 1990. "Urine-Marking in Female Wolves (*Canis lupus*): An Indicator of Dominance Status and Reproductive State." In *Chemical Signals in Vertebrates 5*, edited by D. W. MacDonald, D. Müller-Schwarze, and S. E. Natynczuk, 346–400. Oxford: Oxford University Press.

Sand, H., C. Wikenros, P. Wabakken, and O. Liberg. 2006. "Effects of Hunting Group Size, Snow Depth and Age on the Success of Wolves Hunting Moose." *Animal Behaviour* 72:781e789. https://doi.org/10.1016/j.anbehav.2005.11.030.

Schenkel, R. 1947. "Ausdrucks-studien an Wolfen; Gefangenschafts-Beobachtungen" [Expression studies in the wolf; captivity-observations]. *Behaviour* 1:81–129. Translation from German by Agnes Klasson. https://archive.org/details/SchenkelCaptiveWolfStudy.compressed.

Schmidt, N. M., F. M. van Beest, J. B. Mosbacher, M. Stelvig, L. H. Hansen, J. Nabe-Nielsen, and C. Grøndahl. 2016. "Ungulate Movement in an Extreme Seasonal Environment: Year-Round Movement Patterns of High-Arctic Muskoxen." *Wildlife Biology* 22:253–67. https://doi.org/10.2981/wlb.00219.

Schonberner, D. 1965. "Observations on the Reproductive Biology of the Wolf." *Zeitschrift für Säugetierkunde* 30:171–78.

Schünemann, B., J. Keller, H. Rakoczy, T. Behne, and J. Bräuer. 2021. "Dogs Distinguish Human Intentional and Unintentional Action." *Scientific Reports* 11:14967. https://doi.org/10.1038/s41598-021-94374-3.

Schweizer, R. M., J. Robinson, R. Harrigan, P. Silva, M. Galverni, M. Musiani, R. E. Green, J. Novembre, and R. K. Wayne. 2016. "Targeted Capture and Resequencing of 1040 Genes Reveal Environmentally Driven Functional Variation in Grey Wolves." *Molecular Genetics* 25:357–79. https://doi.org/10.1111/mec.13467.

Schweizer R. M., B. M. vonHoldt, R. J. Harrigan, J. C. Knowles, M. Musiani, D. Coltman, J. Novembre, and R. Wayne. 2016. "Genetic Subdivision and Candidate Genes under Selection in North American Grey Wolves." *Molecular Ecology* 25:380–402. https://doi.org/10.1111/mec.13364.

Seip, D. R. 1992. "Factors Limiting Woodland Caribou Populations and Their Interrelationships with Wolves and Moose in Southeastern British Columbia." *Canadian Journal of Zoology* 70:1494–1503. https://doi.org/10.1139/z92-206.

Serrouya, R., B. N. McLellan, H. van Oort, G. Mowat, and S. Boutin. 2017. "Experimental Moose Reduction Lowers Wolf Density and Stops Decline of Endangered Caribou." *PeerJ* 5:e3736. https://doi.org/10.7717/peerj.3736.

Sinding M-H. S., S. Gopalakrishan, F. G. Vieira, J. A. Samaniego Castruita, K. Raundrup, M. P. H. Jørgensen, M. Meldgaard. B. Petersen, T. Sicheritz-Ponten, J. B. Mikkelsen, et al. 2018. "Population Genomics of Grey Wolves and Wolf-Like Canids in North America." *PLoS Genetetics* 14:e1007745. https://doi.org/10.1371/journal.pgen.1007745.

Smith, D., T. J. Meier, E. Geffen, L. D. Mech, L. G. Adams, J. W. Burch, and R. K. Wayne. 1997. "Is Incest Common in Gray Wolf Packs?" *Behavioral Ecology* 8:384–91. https://doi.org/10.1093/beheco/8.4.384.

Smith, D. W., M. C. Metz, K. A. Cassidy, E. Stahler, R. T. McIntyre, E. S. Almberg, and D. R. Stahler. 2015. "Infanticide in Wolves: Seasonality of Mortalities and Attacks at Dens Support Evolution of Territoriality." *Journal of Mammalogy* 96:1174–83. https://doi.org/10.1093/jmammal/gyv125.

Smith J. E., C. Fichtel, R. K. Holmes, P. M. Kappeler, M. van Vugt, and A. V. Jaeggi. 2022. "Sex Bias in Intergroup Conflict and Collective Movements among Social Mammals: Male Warriors and Female Guides." *Philosophical Transactions of the Royal Society B: Biology* 377:20210142. https://doi.org/10.1098/rstb.2021.0142.

Stahler, D. R., D. R. MacNulty, R. K. Wayne, B. vonHoldt, and D. W. Smith. 2013. "The Adaptive Value of Morphological, Behavioural and Life-History Traits in Reproductive Female Wolves." *Journal of Animal Ecology* 82:222–34. https://doi.org/10.1111/j.1365-2656.2012.02039.x.

Stahler, D. R., D. W. Smith, K. A. Cassidy, E. E. Stahler, M. C. Metz, R. McIntyre, and D. R. MacNulty. 2020. "Ecology of Family Dynamics in Yellowstone Wolf Packs." In *Yellowstone Wolves: Science and Discovery in the World's First National Park*, edited by D. W. Smith, D. R. Stahler, and D. R. MacNulty, 42–60. Chicago: University of Chicago Press.

Taton, N. 2023. "Factors Affecting Gray Wolf (*Canis lupus*) Homesite Selection in Yellowstone National Park." MS thesis, University of Minnesota, St. Paul.

Tener, J. S. 1954. "A Preliminary Study of the Musk-Oxen of Fosheim Peninsula, Ellesmere Island, NWT." *Wildlife Management Bulletin* 1, no. 9. Ottawa: Canadian Wildlife Service.

Tener, J. S. 1965. *Muskoxen in Canada: A Biological and Taxonomic Review*. Ottawa: Queen's Printer.

Theuerkauf, J., W. Jędrzejewski, K. Schmidt, H. Okarma, I., Ruczyński, S. Śiezko, and R. Gula. 2003. "Daily Patterns and Duration of Wolf Activity in the Białowieza Forest, Poland." *Journal of Mammalogy* 84:243–53.

Thiel, R. P., and P. D. DeWitt. 2022. "Territorial Scent-Marking and Proestrus in a Recolonizing Wild Gray Wolf (*Canis lupus*) Population in Central Wisconsin." *Canadian Field-Naturalist* 136:254–61. https://doi.org/10.22621/cfn.v136i3.2907.

Thiel, R. P., W. H. Hall, and R. N. Schultz. 1997. "Early Den Digging by Wolves, *Canis lupus*, in Wisconsin." *Canadian Field-Naturalist* 111:481–82.

Thiel, R. P., A. C. Thiel, and M. Strozewski, eds. 2013. *Wild Wolves We Have Known*. Minneapolis: International Wolf Center.

Townsend, S. E. 1996. "The Role of Social Cognition in Feeding, Marking, and Caching in Captive Wolves, *Canis lupus lycaon* and *Canis lupus bailyei*." PhD diss., University of Colorado, Boulder.

Troelsen, J. C. 1950. *Contributions to the Geology of Northwest Greenland, Ellesmere Island and Axel Heiberg Island* [*Meddelelser om Grønland, udg. af Kommissionen for videnskabelige undersøgelser i Grønland*]. Copenhagen, Denmark: C. A. Reitzel.

Vanderschuren, L. J. M. J., E. J. M. Achterberg, and V. Trezza, 2016. "The Neurobiology of Social Play and Its Rewarding Value in Rats." *Neuroscience & Biobehavioral Reviews* 70:86–105. https://doi.org/10.1016/j.neubiorev.2016.07.025.

Vetter, S. G., L. Rangheard, L. Schaidl, K. Kotrschal, and F. Range. 2023. "Observational Spatial Memory in Wolves and Dogs." *PLoS ONE* 18, no. 9: e0290547. https://doi.org/10.1371/journal.pone.0290547.

Wabakken P., H. Sand, I. Kojola, B. Zimmermann, J. M. Arnemo, H. C. Pedersen, and O. Liberg. 2007. "Multistage, Long-Range Natal Dispersal by a Global Positioning System–Collared Scandinavian Wolf." *Journal of Wildlife Management* 71:1631–34. https://doi.org/10.2193/2006-222.

Ward Jones, M. K., and W. H. Pollard. 2021. "Daily Field Observations of Retrogressive Thaw Slump Dynamics in the Canadian High Arctic." *Arctic* 74:339–54. https://doi.org/10.14430/arctic73377.

Whiten, A. 1991. *Natural Theories of Mind: Evolution, Development and Simulation of Everyday Mindreading*. Oxford: Basil Blackwell, 1991.

Wilmers, C. C., K. Ram, F. G. R. Watson, P. J. White, D. W. Smith, and T. Levi. 2013. "Climate and Vegetation Phenology." In *Yellowstone's Wildlife in Transition*, edited by P. J. White, R. A. Garrott, and G. E. Plumb, 147–63. Cambridge, MA: Harvard University Press. https://doi.org/10.2307/j.ctt2jbx4v.12.

Wilmers, C., M. Metz, D. R. Stahler, M. T. Kohl, C. Geremia, and D. W. Smith. 2020. "How Climate Impacts the Composition of Wolf-Killed Elk in Northern Yellowstone National Park." *Journal of Animal Ecology* 89:1511–19. https://doi.org/10.1111/1365-2656.13200.

World Wildlife Fund. 2023. "Climate Change." https://www.arcticwwf.org/threats/climate-change/.

Young, S. P., and E. A. Goldman. 1944. *The Wolves of North America*. Washington, DC: American Wildlife Institute.

Zub, K., J. Theuerkauf, W. Jędrzejewski, B. Jędrzejewska, K. Schmidt, and R. Kowalczyk. 2003. "Wolf Pack Territory Marking in the Bialowieza Primeval Forest (Poland)." *Behaviour* 140:635–48. https://brill.com/view/journals/beh/140/5/article-p635_4.xml.

Index of Names

Page numbers in italics refer to figures and tables.

Achterberg, E. J. M., 106
Adams, L., 35, 62, 117, 125, *131*, 208
Allen, J. J., 192, 210
Anderson, M., xiv–xv, 82, 102, 124, 130, 135, 142, 147–48, 150, 197, *203*, 203–8, *225*, *plate 4*
Anions, D. W., 135
Apps, C. D., 148
Arthur, S. M., 154
Asa, C. S., 184, 191–94
Audlaluk, L., 19
Ausband, D. E., 63

Ballard, W. B., 171
Barber-Meyer, S. M., 156, 208
Barry, S. J., 70, 211
Bekoff, M., 106, 192
Best, T. L., 136–37
Boitani, L., ix–xi, 35, 44–45, 102, 193, 214–15, 219
Bonnyman, S. G., 136
Boutin, S., 196
Bowen, D., 19
Boyd, D., 35, 63
Brandenburg, J., xiv
Breining, G., 35, 122
Brown, L. C., 213
Brown, R. E., 191–92
Bush, D., 154

Campbell, M. W., 148
Carbyn, L. N., 135
Carmichael, L. E., 59, 86
Caron-Carrier, J., 136–37
Cassidy, K. A., 164, 169, 205, 207
Chapron, G., xv
Charnov, E. L., 153
Ciucci, P., 60, 89, 120, 141, 144
Clark, K., 96, 98, 103–5, 109–10, 115, 147, 168, 170
Cluff, D., xiv–xv, 35, 70, 73–82, 87–88, 94–95, 102, 128, 130, 147, 151, 153, 169, 188, 201–6, *plate 10*, *plate 12*
Crabtree, R. L., 106, 192
Crisler, L., 184

Dalton, T., 32, 51
Dauginis, A. L. A., 213
de Guinea, M., 155
DelGuidice, G. D., 69
de Miguel, F. J., 192
Demma, D. J., 154
DeWitt, P. D., 192
Dick, L., 19
Drummond, J. R., 212
Duck, T. J., 212

Fentress, J. C., 183
Fitts-Cochran, J., 102, 208
Font, E., 184
Frenzel, L. D., 210
Frévol, S., 59, 86, 213
Fritts, S. H., 105, 214
Frost, G. V., 213
Fuller, T., 102, 208
Fuller, W., 207

Gates, V., 35, 73
Gedguadas, N., 35
Gibson, N., 35, 189, *plate 14*
Gipson, P. S., 76
Gitay, H., 211
Gloveli, N., 106
Goldman, H., 103
Goldman, J. G., 155
Goodmann, P. A., 45, 103, 192, 194
Grace, E., 5, 23, 216
Gray, D. R., 127, 129–31, 134, 136
Greeley, A. W., 20
Gregory, J., 211
Gunn, A., 211

Haber, G., 23, 135
Hamilton, W. D., 138, 164, 173–74
Harrington, F., 35, 121–22, 184, 191, 193–94
Harten, L., 155
Hayes, R. D., 148
Heard, D. C., 127
Hearn, B. J., 144
Henry, T. H., 121, 136–37
Hertel, H., 86
Hervieux, D. M., 148
Holt, R. D., 147–48
Hone, E., 127
Howell, S. E. L., 212
Hubert, B., 126
Hutchinson, J., 35
Hutt, C., 35

Janssens, L., xiii, 58
Jędrzejewski, W., 153
Jenkins, D., 147–48
Johnson, C., 35, 181
Joslin, P. W. B., 184

Karl, T. R., 211
Karns, P. D., 210
Kelsall, J. P., 207
Kim, Y. H., 212
Kleiman, D., 192
Klinghammer, E., 45, 103, 192, 194
Kovtun, E., 84
Kreeger, T. J., 98

Lai, S., 134
Laikre, L., 59
Latham, A. D. M., 196
Latour, P. B., 126
Lebovsky, T., 35
Lent, P., 126, 129
Lesins, G., 212
Leutgeb, S., 154
Linnell, J. D. C., 84
Lonardo, L., 40

MacNulty, D., xiv–xv, 35, 56, 79–82, 126, 129–32, 135, 141–42, 156, 162, 168, 203, 206
Magoun, A., 118

Mallory, C. D., 148
Marquard-Petersen, U., 7, 102
Marquez, R., 184
Martin, H., 156
Maule, M., 35, 40–43
McCullough, D. R., 69
McIntyre, R., 40, 54, 59, 117, 169–70, 185, 191, 193
McLellan, B. N., 148
McLeod, P. J., 196
Mech, L. D., ix–xi, xv, 35, 55, *212, plate 10*; "Alpha Status, Dominance, and Division of Labor in Wolf Packs," 159, 174; *The Arctic Wolf: Living with the Pack*, 1; *The Wolf: The Ecology and Behavior of an Endangered Species*, xiv, 63; *Wolves: Behavior, Ecology and Conservation*, 215; *Wolves on the Hunt*, 129, 134
Medwid, W., 35, 72, 87
Merrill, S. B., 39, 58, 75, 113, 150, 155
Mertl-Millhollen, A. S., 192, 194
Miller, F. L., 7, 70, 102, 205, 211
Miller, K., 17
Murie, A., 1, 12, 15, 23, 84, 96, 104, 109, 116–17, 151, 167, 184–85, 188
Muro, C., 162

Nelson, M. E., 127, 135, 138
Nicholls, N., 211
Novakowski, N. S., 207
Nunez, I. B., 192

Oosenbrug, S. M., 135
Ortiz, M., 35
Ortiz-Leal, I., 192

Packard, J. M., 23, 35, 39, 56, 63, 99, 103, 109–17, 169, 171
Palacios, V., 80, 184
Pappas, S., 63
Paquet, P. C., 192, 196
Parker, G. R., 138–39
Parmelee, D., 6–7, 15
Pasitschniak-Arts, M., 28
Paul, W. J., 105
Peary, R. E., 20
Peters, R., 122, 154–55, 191–92, 194
Peterson, R., 35–36, 42–43, 60, 94, 120, 134–41, 169, 210, 219
Pollard, W. H., 213
Post, E., 213
Pulliainen, E., 104

Rantanen, M., 212
Rausch, R. A., 207
Ream, B., 35, 99, 103, 109
Reintjes, F. D., 102, 205
Richard (filmmaker), 189
Robinson, J., 59, 209
Rothman, R. J., 191–92
Rouart, I., 84
Ryman, N., 59
Ryon, J., 183, 191–92

Sand, H., 135
Sanders, J., 35
Schenkel, R., 63, 184
Schmidt, N. M., 126
Schonberner, D., 103, 110
Schüneman, B., 40
Schweizer, R. M., 59, 209–10
Seal, U. S., 154
Seip, D. R., 148
Serrouya, R., 148
Sinding, M.-H. S., 59, 209
Smith, Deborah, 59
Smith, Douglas, 56, 96, 126, 129, 132, 135, 145, 156, 162, 168, 196, 201
Smith, J., 164
Spaulding, T., 35
Stahler, E., 63, 164, 169, 193
Stephenson, R. O., 208
Sternal, Ron, 35

Strozewski, M., 84
Swain, Una, 35

Taton, N., 89, 144
Taylor, M. E., 241
Tener, J. S., 124, 126–27, 132, 135, 142
Thiel, A. C., 84
Thiel, R. P., 84, 97, 192
Townsend, S. E., 117, 121
Trezza, V., 106
Troelsen, J. C., 7
Turner, J., 32, 35, 57

Ullrey, D. E., 69

Vanderschuren, L. J. M. J., 106
Vetter, S. G., 121
Viranyi, Z., 40–41
Vnukovsky, A., 96
vonHoldt, B. M., 59, 209–10

Wabakken, P., 62
Ward Jones, M. K., 213
Whiten, A., 40
Wilmers, C. C., 210
Wolf, P. C., 56, 114, 117, 171

Young, S., 103

Zub, K., 191–92

Index of Subjects

Page numbers in italics refer to figures and tables.

acclimation, 9
aggression: behavioral studies and, 46, 63, 84, 111–12, 132, 168, 172–73, 194–95, 199; den studies and, 36, 63; muskoxen and, 132; pups and, 111–12, 199; territoriality and, 194–95, 199
airplanes, 8–9, 27, 65–66
Alaska: behavioral studies and, 171, 207–8, 214; caches and, 118; Denali National Park, 12, 23, 36, 39, 42, 53, 67, 75, 116–17, 147, 151, 153, 175, 196, 215; den studies and, 2, 16, 19, 36; Ellesmere Island and, 5, 16, 19; hunting and, 153; Queen Elizabeth Islands and, 5; radio collars and, 12, 39, 53, 75
Alert military base, 4, 6, 19–20, 137
all-terrain vehicles. *See* ATVs
alpha females, 63, 173
alpha males: behavioral studies and, 173; den studies and, 63; headquartering and, 176, 181, 183; hunting and, 158–59; territoriality and, 199
"Alpha Status, Dominance, and Division of Labor in Wolf Packs" (Mech), 159, 174
ambushing, 141, 160, 162, 173, 216
apparent competition, 147–48
Arctic (journal), 7
arctic fox (*Vulpes lagopus*), 227
arctic hares (*Lepus arcticus*), *plate 5*; appearance of, 136–38; attacks on, 138, 147; behavioral studies and, 62, 86; breeding females and, 140; breeding males and, 142; caches, 116, 119–20, 179; counts of, 68; den studies and, 27, 56, 62; distribution of, 136; Eureka Weather Station and, 68, 124, *138*, *146*, 147; Explorer and, 142; food and, 140–48; herd size of, 138–40; hunting and, 8, 27, 54–58, 62, 105, 129, 137–45, 148, 165–67, 173, 178, 186, 216, *plate 16*; leverets, 56, 62, 86, 99, 104–5, 116, 119–20, 124, 137–43, 166–67, 173, 178–79, 186, 190; litter size and, 137; Mom and, 140–41; mothers and, 138; Nunavut and, *143*, *145*, 148; nursing

and, 138–39, 144; pups and, 99, 104–5; radio collars and, 137; range size of, 137; sleep and, 140–41; Slidre Fiord and, 68, 144; speed of, 136–37; survival and, 67–68, 139–40, 216; tundra and, 136–37, 139; vegetation and, 136, 138; weight of, 136; Whitey and, 139; yearlings and, 105, 141; zigzag defense of, 137, 140
Arctic Ocean, 44–45, 212
arctic tern (*Sterna paradisaea*), 17, 227
arctic willow (*Salix arctica*), 18, 136, 227
Arctic Wolf, The: Living with the Pack (Mech), 1
arousal, 27, 194
Astrolab Peak, 147
Atmospheric Environment Services, 18
attacks: arctic hares and, 138, 147; caribou and, 147; congregating leverets and, 138; on dogs, 216; growling before, 55; on humans, 84, 88; moose and, 84; muskoxen and, 41, 62, 125–26, 130–34, 147, 159; pup play and, 105–6; territoriality and, 194; trespassers and, 11, 52, 194, 207
ATVs, *plate 10*; arctic hare studies and, 140; behavioral studies and, 66–75, 87–88, 165, 215; darting from, 81; den studies and, 3, 5, 9–12, 18, 22–27, 32–36, 40–45, 51, 54, 57, 94–95; funding and, 9, 32; headquartering and, 175, 178; howling observations and, 189; hunting studies and, 151, 156; importance of, 24; territoriality studies and, 196; wolves' fear of, 5, 25
Axel Heiberg Island: caribou and, 148; Fosheim Peninsula and, 17, 78, 148, 203, 205, 207; Gibbs Fiord, *102*; location of, *4*, *16*; Princess Margaret Range and, 211; radio collars and, *223*, *225*; range size and, *226*; territoriality and, 201, 203, 205; Troelsen on, 7
Aylmer Lake, 70, 129

babysitters, 15, 26, 51, 150
backtracking, 74, 117, 161
Baffin Island, 98, 103–4, 115, 147, 168
bait buckets, 51
bark-howl, 58, 168, 184
barking, 1, 10, 13, 58, 93, 168, 171, 184
Bathurst Island, 2, 127, 148
Bauman Fiord, 205
beavers (*Castor canadensis*), 17, 215, 227
behavioral studies: aggression, 46, 63, 84, 111–12, 132, 168, 172–73, 194–95, 199; Alaska and, 171, 207–8, 214; alpha females, 63, 173; alpha males, 63, 158–59, 173, 176, 181, 183, 199; "Alpha Status, Dominance, and Division of Labor in Wolf Packs," 159, 174; arctic hares and, 62, 86; arousal, 27, 194; barking, 1, 10, 13, 58, 93, 168, 171, 184; bigger picture and, 206–8; breeding females, 86, 167–71, 215–16; breeding males, 72, 86, 166–74; Brutus, 70–73, 85–88; Denali National Park and, 67, 215; documentaries and, 84–88; dominance, 168–71; Eureka Weather Station and, 68–73, 88, 216; Explorer, 64, 85–86, 170; food and, 66, 164–67, 171–74, 208; Fosheim Peninsula and, 206–8, 216; foxes, 71–72; greetings, 2, 46, 201; High Arctic and, 67, 72, 87–88; howling, 86–88, 93, 167–68, 184–90, 215–16; hugging, 45–46; hunting, 9, 23, 42, 54–55, 67, 73, 86, 150–69, 173, 178, 207, 214–16; inclusive-fitness theory and, 173–74; "licking up" action, 46, 49, 111–12, 165–67, 170, 173, 180, 182–83, *plate 17*; migration, 39, 137, 144, 213; Mom,

85, 173; mothers, 85, 165, 167, 171–72, *plate 6*; muskoxen, 67–73, 86–87; *National Geographic* and, 66; nursing, 64, 70, 72, 167, 171–72, 216; playing, *100*, 105–7; pups, 64–65, 68–72, *plate 18*; radio collars and, 65, 88, 206–7, 215–16; regurgitation, 165–67, 173; rendezvous sites, 71–73; roles, 164–74; sleep, 86; snarling, 55, 173; status, 174; submission, 48–49, 159, 170–73, 180; survival and, 164; tracking, 70; tundra and, 86–88; Whitey, 46, 85–86, 165–73; *Wild Wolves We Have Known*, 84–88; yearlings and, 85, 170–73; Yellowstone and, 169–70, 173, 208, 214

beluga whales (*Delphinapterus leucas*), xi, 19, 227

Białowiez Forest, 177

birthing: dangers after, 57; dens for, 44, 47, 97; estimating date of, 98, 103, 105; pups and, 44, 47, 57, 97, 98, 103, 105, 139; timing of, 116

bison (*Bison bison*), xv, 135, 227

black bears (*Ursus americanus*), xv, 227

Blackmask, 47–50, 92, 123, 197

Blacktop Ridge, 4, 18, 40, 42, 68, *146*, 147

blizzards, 82, 215

blood samples, 11, 31, 83

blowguns, 75–76

bonded pairs, 70, 192

books, ix, 1, 6, 7, 8, 12, 15, 26, 30, 63, 84, 129, 134, 175, 183, 190, 215, 216, 217. *See also individual titles*

boxing, 46

breeding females: alpha females and, 63, 173; arctic hares and, 140; behavioral studies and, 86, 167–71, 215–16; bonded pairs and, 70, 192; caches and, 115–18; den studies and, 23–26, 38–43, *45*, 51, 53, 56–63, 90; headquartering and, 178, 180; howling and, 188; hunting and, 150–55; lactation and, 3; mating and, 63, 103; nursing and, 5, 39–40, 53, 57, 70, 72, 95, 99, 104, 109–12, 138–39, 144, 171–72, 179, 186, 216; pups and, 98, 107 (*see also* pups); Slidre Fiord and, 76–77; territoriality and, 192, 194, 204

breeding males: alpha males and, 63, 148–49, 173, 176, 181, 183, 199; behavioral studies and, 72, 86, 166–74; caches and, 117, 123; den studies and, 38, 41–44, *45*, 56–58; hares and, 142; headquartering and, 176, 181–82; howling and, 186–201; hunting and, 153, 159; inclusive-fitness theory and, 173–74; pups and, 105–7, 112–15; territoriality and, 192

British Broadcasting Company (BBC), 32

browning, 213

Brutus: behavioral studies and, 70–73, 85–88; boldness of, 86–88; caribou and, 147; darting of, 76; death of, 80; Eureka Weather Station and, 70–77, 88, 95–96, 201, 203; hunting and, 151; muskoxen and, 130; radio collars and, 75–76, 88, 201, 203; Slidre Fiord and, 74–83; territoriality and, 201–7; toys and, 107–8

caches: Alaska and, 118; arctic hares, 116, 119–20, 179; breeding females and, 115–18; breeding males and, 117, 123; caribou, 116; Denali National Park and, 116–17; dogs and, 116; dominance and, 119, 122; Explorer and, 118–20; food, 30, 41, 56, 58, 62, *100*, 105, 108, 112, 116–23, 137–38, 158–59, 162, 166, 171, 179–83, 191, 198, 215; fuel, 18, 34, 159; hunting and, 120; locations of, 117; marking, 121–23; Mom and, 122; muskoxen, 116, 118, 123, 166; pups and, 116, 118;

caches: Alaska (*continued*)
regurgitation and, 117, 123; retrieval and, 118–23; scent and, 117, 121; seals, 179–80; secrecy of, 117–18; stealing from, 117–18; tracking and, 117; urination and, 121–22; Whitey and, 118, 122–23; yearlings and, 119–20
calves, 27–30, 41, 42, 56, *67*, 69, 87, 116, 117, 124, 126, 130–35, 145–47, 152, 156, 157–60, 177, 215
camels (*Camelus* sp.), 17, 227
cameras, 22, 28, 33, 157, 175, 215
Canadian Arctic Archipelago, 5, 102
Canadian Field Naturalist (journal), 124–25
cancer, 80, 88
cannibalism, 60–61
Cañon Fiord, *4*, 17, *102*, *223*, *225–26*
captive colonies (of wolves), 2, 63, 192
captivity, 21, 190
caribou (*Rangifer tarandus*), xiv, 227; attacks on, 147; Axel Heiberg Island and, 148; Brutus and, 147; caches and, 116; calves of, 116, *146*, 147, 215; Denali National Park and, 147; Ellesmere Island and, 6, 82; Eureka Weather Station and, 6, *146*, 147; food and, 17, 116, 145, 147–48; Fosheim Peninsula and, 145, 148; High Arctic and, 145; hunting, 19, 39, 54, 62, 81–82, 124, 135–36, 145–49, 206, 211, 215; migration of, 39; Peary (*Rangifer tarandus pearyi*), 6, 17, 82, 124, 136, 145, 148, 206, 211, 228, *plate 4*
Clearwater, 42
climate change, 209–13
clusters, kill or location. *See* kill cluster
coat characteristics, 26
cognitive maps, 154–55
collar. *See* radio collars
collared lemming (*Dicrostonyx groenlandicus*), 17, 105, 108, 115, 123, 227
Committee on the Status of Endangered Wildlife in Canada (COSEWIC), 145, 148
Cornwallis Island, 33–34, 65
coyotes, x, 121, 192

danger, 5, 7, 46–47, 78
darting, 11, 75–78, 81, 83, 88
daylight, 6, 9, 17, 23, 78
defecation, 156, 193–94
Denali National Park, 16; behavioral studies and, 67, 215; caches and, 116–17; caribou and, 147; Clearwater pack and, 42; den studies and, 23, 36, 39, 42, 53; headquartering and, 175; hunting and, 151, 153; Murie and, 151; radio collars and, 12, 39, 53, 75; research at, 12; territoriality and, 196
dens: activity at, 2, 15, 42, 46, 54, 58, 99, 105, 107, 111–12, 113–14, 118, 150–51, 167–68, 175–80, 182–83, 185–86; finding, 1–3, 5, 7, 10–11, 18, 40–42, 43, 62, 70–71, 74, 77, 90, 95; fiord den, 47, 51, 90, 94, 198; fox dens, 71, 72, 77, 90, 94, 95, 97; location in territory, 89, 144, 196; pup emergence from, 99, 103–4; repeated use, 23, 90; rock-cave den, 22–23, 44, 51, 56, 71, 72, 90, 94, 96, 109, 144, 225; visiting outside denning season, 81
density: of caribou, 147; of prey, 17; of wolves, 127, 208
den studies: aggression, 36, 63; Alaska and, 16, 19, 36; alpha males, 63; arctic hares, 27, 56, 62; birthing, 44, 47, *97*; breeding females, 23–26, 38–43, *45*, 51, 53, 56–63, 90; breeding males, 38, 41–44, *45*, 56–58; Brutus and, 95–96; caves and, 22–23, 44, 47, 50, 71, 89–90, 94, 96, 99, 109, 207–8; Denali National Park and, 23, 36, 39, 42, 53; Ellesmere Island habitat and,

1–9; Eureka Weather Station and, 18–21, 24, 33–36, 40, 43, 47, 53–54, 58, 60–62, 95–96; Explorer and, 55, 61–62; food, 12, 15–19, 25, 27, 33–36, 39, 41–42, 51–52, 60–61, 89, 94, 97; Grace and, 5; headquartering, 175–83; High Arctic and, 34, *50*, 59, 62; howling, 1, 10–13, 55, 58; hunting, 14, 19, 23–27, 35, 39, 41–42, 50, 53–58, 97; Inuit people and, 19, 34, 62; living with pack and, 21–30; Mom, 7, 11–15, 40–51, 59, *92*; mothers, 5, 15, 39, 43, 53, 56, *99–100*, 104, 109–11, 151, 165, 167; muskoxen interactions, 12, 17, 19, 26–30, 41–42, 51, 54, 56–57, 62, 94; *National Geographic* and, 18, 21, 23, 26, 32–34; nursing, 5, 15, 39–40, 48–49, 53, 57, 91–95, 186; observational data on, *91–94*, *222–26*; playing, 9, 46, 94, 140; pups, 1–7, 10–15, 20–27, 30, 35, 38–44, 47–49, 53–62, 77–83, 89–91, 94–118, 130, 140, 144, 149–52, 159–60, 165–68, 171, 175, 178–86, 196–208, 215, *plates 1–3*, *6–9*, *14–15*, *17–19*; radio collars and, 11–12, 39, 43, 53; range size, 39–40, *226*; regurgitation, 26, 39, 49, 58; rendezvous sites, 42–47, 54, 56, 60, 95–97; Resolute Bay and, 6, 33–34, 62, 65–66; sleep, 13, 19, 24–25, 27, 41–42, 49, 55, 57, 96, *100*, 105, 109, 140–41, 196; Slidre Fiord and, 47, 90, 94–95; status, 38, 46–47, 63; tracking and, 12, 21, 31, 39; traps and, 2, 11, 21, 31, 95; tundra and, 13, 18, 27–30, 36, 40, 95, 97, *plate 2*; urination, 35, 46; vegetation, 15, 17–19, 57, 96; Whitey, 38, 42–51, 54–61, 90–94; yearlings, 10, 15, 38, 41–43, 54–55, 61
Department of Defense, 61
Department of Environment, 206–7
desert, 7, 16, 178
Devon Island, 126
dispersal, 59, 62, 86, 144, 212; of collared wolves, 216
DNA, 59, 61, 164
documentaries: behavioral studies and, 84–88; Dalton and, 32, 51; Ellesmere Island and, 26, 32, 51, 82, 152, 205, 217; funding and, 82, 83; *National Geographic* and, 26, 32, 152; Slidre Fiord and, 205. *See also specific films*
dog food, 25, 51, 123
dogs, x; attacks on, 216; caches and, 116; domestic, 40; ground scratching and, 194; pet, 21; playing and, 106; sled, 216; sleep and, 177
dominance: behavioral studies and, 168–71; caches and, 119, 122; den studies and, 1–7, 10–15, 20–27, 30, 35, 38–44, 47–49, 53–62; hierarchy and, 63; hunting and, 159; roles and, 168–71; “standing over” posture and, 111; status and, 63 (*see also* status); territoriality and, 191–92
drivers of dispersal, 11, 144
drones, 22, 33
drought, 210
dump. *See* landfills

Eastwind Lake, 18, 153, 159
ecosystem issues, 62, 84, 210, 213
elk, 135, 141, 155–56, 174, 208, 214–15
Ellesmere Island, x–xi; Alaska and, 16, 19; behavioral studies and, 64; caribou, 6, 82; den studies and, 11 (*see also* den studies); description of, xiii; documentaries and, 26, 32, 51, 82, 152, 205, 217; Eastwind Lake, 18, 153, 159; Eureka Weather Station, *4*, 6, 9, 18 (*see also* Eureka Weather Station); Fosheim Peninsula and, 17–20, *102*, 124, 204–8; Greenland proximity of, 5, 7, 16, 19, 136; habitat of, x, xiii–xiv,

Ellesmere Island (*continued*)
1–9; High Arctic and, 1, 6–7; logistics of getting to, 6, 65–66; muskoxen, 6–8; *National Geographic* and, 5–6; North Pole and, 1, 3, 15–16, 18, 20, 66, *226*; Nunavut and, 1, 3–4, *45*, *69*, *91*, *97*, *99*, *101*, *102*, *113*, *114*, *152*, *200*, *203*, *204*; observation data, *222–26*; Qikiqtani Truth Commission and, 19; Romulus Lake, 18; terrain and, 13; travel expenses to, 6; tropical past of, 17–18; tundra and, 3, 5, *plate 2*; uniqueness of wolves, xiii–xv; vegetation and, 17; weather of, 16–17
Endangered Species List, 215
ermine (*Mustela erminea*), 17, 227
estrous, 81, 98
Eureka Weather Station: airstrip of, 95, 178; arctic hares and, 68, 124, *138*, *143*, 147; Atmospheric Environment Services and, 18; behavioral studies and, 68–73, 88, 216; Brutus and, 70–77, 88, 95–96, 201, 203; building of, 18; caribou and, 6, *146*, 147; climate change and, 212; den studies and, 18–21, 24, 33–36, 40, 43, 47, 53–54, 58, 60–62, 95–96; Fosheim Peninsula, 18, 124; headquartering and, 74, 178; landfill of, 33, 107, 166, 194, 199; location of, 4; muskoxen and, 124; research methodology and, *222*; roadway from, 68; sled dogs and, 216; Slidre Fiord and, 9, 74, 78; temperature recordings at, 212; territoriality and, 194, 197–201, 204; toys from, 107; wolf observations and, 7, 10, 20–21, 33, 40, 60–62, 68–72, 88, 96, 147, 166, 178, 194, 197–98, 216
evolution, 83, 106, 140, 164–165, 171, 172, 174, 214
experimentation, 14, 41, 106, 121, 165, 166, 171–173, 185, 216. *See also* tests, of wolves
Explorer: arctic hares and, 142; behavioral studies and, 64, 85–86, 170; caches and, 118–20; den studies and, 55, 61–62; headquartering and, 180–83; hunting and, 159; muskoxen and, 126; pups and, 105, 107
extinguishing (a learned response), 51–52

feeding, by humans, 7, 9, 10, 24, 25, 38, 48, 51, 60, 61, 70, 87, 117, 154, 165, 171–73, 179, 183
feeding tests, 9, 51, 114, 165, 171–73
Female 5176, 86
fetus resorbing, 57, 65
figures, 2, 3, 4, 8, 28, 50, 67, 110, 128, 138, 143, 145, 158, 176, 200, 202, 203, 204, 212, 223
Fleming College, 96
flying lemurs (*Cynocephalus volans*), 17, 227
food: accepting, 12; arctic hares, 140–48; bait buckets, 51; behavioral studies and, 66, 164–67, 171–74, 208; caches, 30, 41, 56, 58, 62, *100*, 105, 108, 112, 116–23, 137–38, 158–59, 162, 166, 171, 179–83, 191, 198, 215; caribou and, 17, 116, 145, 147–48; climate change and, 213; competition for, 140–41, 172–73; den studies and, 12, 15–19, 25, 27, 33–36, 39, 41–42, 51–52, 60–61, 89, 94, 97; dog, 25, 51, 123; feast-or-famine approach to, 116, 209; headquartering and, 179, 182–83; howling and, 13, 151, 159, 188–89; human leftovers, 7, 9; hunting, 150–63 (*see also* hunting); "licking up" action and, 111–12, *plate 17*; meat scraps, 10, 13–14, 172; muskoxen and, 124, *plate 20*; nursing and, 99 (*see also* nursing); pups and, 7, 15, 25, 27, 39–42, 56, 60, 65, 94, 96, *100*, 103–5, 108, 111–17, 123, 147, 150, 164–67, 171–74, 183, 208; regurgitation, 26, 39, 49 (*see also*

regurgitation); scent and, 7, 25, 117, 121–22, 191, 195; seals and, 51, 145, 148; stalking and, 51–52; starvation, 60–62, 116, 210–11; stealing, 105, 117–18; territoriality and, 191, 195, 197; vegetation, 17, 129, 155, 210, 213
Fort Conger, *4*, 20
Fosheim Peninsula, 65; Axel Heiberg Island and, 17, 78, 148, 203, 205, 207; behavioral studies on, 206–8, 216; caribou and, 145, 148; Ellesmere Island and, 17–20, *102*, 124, 204–8; Eureka Weather Station, 18, 124; location of, *4*, 17; muskoxen and, 124, *128*, 130; number of wolves on, *102*; Slidre Fiord and, 18–19, 78; terrain of, 18; territoriality and, 201–5; weather of, 17
fossils, 17
foxes: behavioral studies and, 71–72; boxing and, 46; as carnivores, 17; dens of, 71–72, 77, 90, 94–97; red, 46, 121; Slidre Fiord and, 77; status and, 46; as toys, 107; trapping, 154
funding, 206; documentaries and, 82–83; finding, 35; Gates and, 73; *National Geographic* and, 26, 32–34, 81

geese, 136
genetics: beginning use of, 12; den studies and, 59, 61, 83, 86; diversity, 59; DNA, 59, 61, 164; reproduction and, 162, 164, 171–72
Gibbs Fiord, *102*
Glacier National Park, 209
glaciers, 6, 16, 45, 82
global warming, 209
Grayback, 38, 106
Grayback II, 142, 181–83
Greely Fiord, *4*, 17, 156
Greenland, *4*; Alert and, 20; Ellesmere Island proximity to, 5, 7, 16, 19, 136; litter sizes in, 102; Resolute Bay and, 34; Thule, xiii, 19, 34, 79
greetings, 2, 46, 201
grid cells, 154–55
Grise Fiord, xv, *4*, 6, 19, 79, 82–83, 207, 216
grizzly bears (*Ursus arctos*), xv, 54, 227
ground scratching, 194–96, 199
growling, 48–50, 55, 112, 190, 199
gyrfalcon (*Falco rusticolus*), 17, 227

headquartering: breeding females and, 178, 180; breeding males and, 176, 181–83; Denali National Park and, 175; Eureka Weather Station and, 178; Explorer and, 180–83; food and, 179, 182–83; hunting and, 176–81; muskoxen and, 177–79, 181–83; nursing and, 179, 183; pups and, 175–83; regurgitation and, 179, 182–83; rendezvous sites and, 181; sleep and, 175–81; Slidre Fiord and, 178; tundra and, 180; Whitey and, 177–83; Yellowstone and, 175
helicopter, 9, 15, 20, 42, 56, 64, 68, 73, 74, 77, 80, 81, 82, 83, 95
hierarchy, 63
High Arctic, x–xi; behavioral studies and, 67, 72, 87–88; caribou and, 145; climate change and, 209–13, 216; den studies and, 34, *50*, 59, 62; Ellesmere Island and, 1, 6–7; Polar Continental Shelf Program (PCSP) and, 34; Queen Elizabeth Islands, xiii, 5, 18, 62, 129, 136, 145
hippocampus, 154
hospitals, 5
howling: bark, 1, 10, 13, 58, *93*, 168, 184; behavioral studies and, 86–88, *93*, 167–68, 184–90, 215–16; breeding females and, 188; breeding males and, 186–201; counting, 187, 190, 195; den studies and, 1, 10–13, 55, 58;

howling: bark (*continued*)
experimentally testing, 185–90; food and, 13, 151, 159, 188–89; howling and, 167–68, 184–90; hunting and, 151, 159, 185–88; Mom and, 184, 189; mothers and, 187; muskoxen and, 187; pups and, *99–100*, 107; silent, 185; territoriality and, 184, 194–98; testing, 188–90; tundra and, 189; Whitey and, 184–90; yearlings and, 187–88; Yellowstone and, 185
Hudson Bay, 1, 16
hugging, 45–46
Hunter Trapper Association (HTA), 79
hunting: actions in daily, 150–63; Alaska and, 153; alpha males and, 158–59; ambushing, 141, 160, 162, 173, 216; arctic hares, 8, 27, 54–58, 62, 105, 129, 137–45, 148, 165–67, 173, 178, 186, 216, *plate 16*; behavioral studies and, 9, 23, 42, 54–55, 67, 73, 86, 150–69, 173, 178, 207, 214–16; beluga whales, 19; breeding females and, 150–55; breeding males and, 153, 159; Brutus and, 151; caches and, 120; caribou, 19, 39, 54, 62, 81–82, 124, 135–36, 145–49, 206, 211, 215; ceremony before, 151; climate change and, 210; cognitive maps and, 154–55; Denali National Park and, 151, 153; den studies and, 14, 19, 23–27, 35, 39, 41–42, 50, 53–58, 97; destinations for, 153; dominance and, 159; Explorer and, 159; headquartering and, 176–81; howling and, 151, 159, 185–88; human accompaniment on, 26; Hunter Trapper Association (HTA) and, 79; Iviq Hunters and Trappers Association, 207; methods of killing, 124–27, 132–33, 157–59; Mom and, 140–41, 159; mothers and, 151; muskoxen, 8, 19, 26–30, 41, 54, 69–70, 73, 86, 105, 124–36, 139–44, 147–53, 156–62, 169–72, 176–83, 187, 194–95, 199, 207–11, 215, *plate 11*, *plate 13*; *National Geographic* and, 152, 159; nursing and, 150; pups and, 112; range of, 152–53; rendezvous sites and, 155, 159; scent and, 156; seals, 19, 124, 136, 145, 148–49, 207; sleep and, 140–41; Slidre Fiord and, 77–80; stalking, 139, 157, 160–62; status and, 159; success rates of, 134–35, 141, 177; territoriality and, 193; timing of, 150–51; traveling speed of, 151–52, 155; tundra and, 86, 151; urination and, 156–58; *Wolves on the Hunt*, 129, 134; yearlings and, 152
hybridization, x

icebergs, 9, 16
ice sheets, 18
Igloolik, 82
igloos, 19
inbreeding, 59, 86, 212
inbreeding depression, 59, 213
inclusive fitness, 173
Industrial Revolution, 214
infanticide, 196, 205
Information Age, 214
International Union for the Conservation of Nature (IUCN), 36, 44, 215
International Wolf Center: behavioral studies of, 70–73, 84, 87, 215–16; den studies of, 34–36; founding of, 34, 36; pup observation and, 103; staff of, 35
intruders. *See* trespassing
Inuit people: Anderson and, 207; den studies and, 19, 34, 62; Grise Fiord and, 4, 6, 19, 79, 82–83, 207, 216; Hunter Trapper Association (HTA) and, 79; PCSP and, 66; radio collars and, 79, 82; relationship with researchers, 79; seals and, 165; Thule people and, 19; transplanting of, 19

Inuit *qaujimajatuqangit*, 219
Inuk people, 19, 80, 82
Inuktitut, 80
Isle Royale National Park, Michigan, 2, 8; behavioral studies and, 84; den studies and, 11, 27, 42; moose and, 27, 42, 84, 132, 134, 160; range size and, 152; travel speeds and, 151
Italy, 2
Iviq Hunters and Trappers Association, 207, 219

jackals, x
jackrabbits, 136
journals, 5, 7, 98

kill, 13, 27, 30, 41, 51, 55, 69, 81, 83, 86, 94, 113, 117–119, 123, 124, 126, 130–31,133–34, 142, 147, 149, 152, 156–59, 181–83, 195
kill cluster, 78, 80, 142, 207
kill rate, 134, 142, 206, 208
Kingdom of the White Wolf (National Geographic), 33
kin-selection, 174

Lake Hazen, 19, 137
landfills, 33, 107, 166, 194, 199
latitude, 10, 16, 96, 98, 103, 208, 209, 226
leadership, 56, 155, 167–70
Left Shoulder, 38, 85–86, 185, 190
leopards (*Panthera pardus*), 227
leverets, 56, 62, 86, 99, 104–5, 116, 119–20, 124, 137–43, 166–67, 179, 186, 190
"licking up" action: food and, 111–12, *plate 17*; submission and, 46, 49, 111–12, 159, 165–67, 170, 173, 180, 182–83
lions (*Panthera leo*), 86, 190, 227
litter size, 59, 102, 115, *plate 19*
livestock, 215–16
live-trapping, 2, 11, 21, 31, 190
long-tailed jaeger (*Stercorarius longicaudus*), 17, 215, 227

magnetic resonance imaging (MRI), 216
Manitoba, 2
marking: bonded pairs and, 70; caches and, 121–23; double, 70; reproduction and, 191; research methodology and, 39; scent, 11, 38, 55, 67, 70, 121–22, 156, 186, 191–96, 199; territoriality and, 11, 67, 70, 121–23, 156, 179, 191–99; urination, 46, 121–22, 156, 192–94; vomeronasal organ (VNO) and, 191–92
mating, 63, 65, 103
meat scraps, 10, 13–14, 24–25
Melville Island, 7, 127
Mid-Back, 157, 159
migration, 39, 137, 144, 213
milk. *See* nursing
Minnesota, 1, 2, 5, 6, 27, 32, 36, 37, 39, 53, 60, 62, 65, 67, 75, 79, 80, 86, 94, 154, 196, 214, 215, 231, 236, 238, 239, 241, 244
Mom: arctic hares and, 140–41; behavioral studies and, 85, 173; caches and, 122; den studies and, 7, 11–15, 40–51, 59, 92; Ellesmere Island and, 7; howling and, 184, 189; hunting and, 140–41, 159; as Nipples, 113–14; pups and, 7, 11–15, 40–44, 47–50, 59, 85, 109–13, 115, 173, 189; status of, 43–47; territoriality and, 196–97; Whitey and, 42–50, 59, 85, 115, 122, 173, 184, 196–97; wounds of, 47–48
moose (*Alces alces*), xv, 227; attacks on, 84; hunting, 132–35, 148, 214; Isle Royale National Park, 27, 42, 84, 132, 134, 160
mortality, 2, 69–70, 82, 96, 208, 211
mothers: arctic hares and, 138; behavioral studies and, 85, 165, 167, 171–72,

mothers: arctic hares and (*continued*) *plate 6*; birthing and, 44, 47, 57, 97, 98, 103, 105, 139; den studies and, 5, 15, 39, 43, 53, 56, *99–100*, 104, 109–11, 151, 165, 167; howling and, 187; hunting and, 151; muskoxen and, 126; nipples and, 3, 5, 53–54, 57, 64, 102, 109, 113–14; nursing and, 5 (*see also* nursing); Parmalee on, 7; regurgitation and, 39, *100*, 104; territoriality and, 192; weaning and, 39, 72, 103, 110–11, 168, 171–73
Mount McKinley, 1, 12, 23
mudflats, 17
muskoxen (*Ovibos moschatus*), xv, 228; aggression and, 132; appearance of, 126; attacks and, 41, 62, 125–26, 130–34, 147, 159; behavioral studies and, 67–73, 86–87; Brutus and, 130; caches and, 116, 118, 123, 166; calves of, 27–30, 41–42, 56, 62, *69*, 87, 116–17, 124, 126, 130–35, 145, 147, 152, 156–60, 177; counts of, 68; danger of, 7; deadly kicks of, 28; den studies and, 12, 17, 19, 26–30, 41–42, 51, 54, 56–57, 62, 94; Ellesmere Island and, 6–8; Eureka Weather Station and, 124; Explorer and, 126; eyesight of, 129; fighting with, 168–69; food and, 124, *plate 20*; Fosheim Peninsula and, 124, *128*, 130; grazing of, 129–30; headquartering and, 177–83; herd density of, 177; herd size and, 127–28; howling and, 187; hunting, 8, 19, 26–30, 41, 54, 69–70, 73, 86, 105, 124–36, 139–44, 147–53, 156–62, 169–72, 176–83, 187, 194–95, 199, 207–11, 215, *plate 11*, *plate 13*; lifespan of, 124; mothers and, 126; Nunavut and, 129, *131*; penis, 116; pups and, 27, 54, 56, 67–68, 72, 81, 86, 94, *100*, 105, 113–16, 123, 130, 143–44, 150, 152, 159, 168, 178–79, 183, 187, 208; radio collars and, 129; range size of, 129; shaggy hair of, 126; Slidre Fiord and, 75, 79, 81–82, 128; solitary, 127; survival and, 67–68; territoriality and, 194–95, 199, 205; tundra and, 130; vegetation and, 129; weight of, 126; yearlings and, 41, 130, *131*, 133–35

Nansen Sound, 17
narwhals (*Monodon monoceros*), 17, 228
National Geographic, xiv; behavioral studies and, 66; den studies and, 18, 21, 23, 26, 32–34; documentaries and, 26, 32, 152; Ellesmere Island and, 5–6; funding and, 26, 32–34, 81; hunting and, 152, 159; *Kingdom of the White Wolf*, 33; Slidre Fiord and, 79, 81; *White Wolf*, 32–33, 87, 147, 152, 159, 199
Nature of Things (CBC), 33
Newfoundland, 136
nipples: competition for, 109; distended, 64; Mom and, 113–14; nursing and, 3, 5, 53–54, 57, 64, 102, 109, 113–14
nomadism, 11, 39, 68, 96, 105, 132, 152, 153, 175
North Pole, x; Ellesmere Island and, 1, 3, 15–20, 66, *226*; Eureka and, 34; Peary and, 20
Northwest Territories, 2, 34, 70, 87, 129, 207
Nunavummiut, 82
Nunavut: arctic hares and, *143*, *145*, 148; Cornwallis Island, 33–34, 65; Ellesmere Island studies, 1, 3–4, *45*, *69*, *97*, *99*, *101–2*, *113–14*, *152*, *200*, *203–4*; Eureka Weather Station, 6 (*see also* Eureka Weather Station); Government of, 79, 82, 129, 148, 206–7; Grise Fiord, 4, 6, 19, 79, 82–83, 207, 216; Igloolik, 82; Inuit Hunter Trapper Association (HTA) and, 79; muskoxen and, 129, *131*

nursing, *plate 7*; arctic hares and, 138–39, 144; behavioral studies and, 64, 70, 72, 167, 171–72, 216; den studies and, 5, 15, 39–40, 48–49, 53, 57, 91–95, 186; headquartering and, 179, 183; hunting and, 150; length of, 110; logistics of, 109–14; nipples and, 3, 5, 53–54, 57, 64, 102, 109, 113–14; number of times, 15; pup growth and, 99, 104, 109–14; simultaneous, 109; termination of, 109–10; weaning and, 39, 72, 103, 110–11, 168, 171–73

oases, 16, 18
officer-in-charge (OIC), 58, 61
Old Man, 61

PBS, 32
Peary caribou (*Rangifer tarandus pearyi*), 17, 228, *plate 4*; climate change and, 211; hunting, 124, 136, 145, 148, 206; Slidre Fiord and, 82
pelage, 22, 39, 145
Perch Lake, 86
permafrost, 23, 36, 89, 212
permits, 57–58, 79–80, 83, 206
photography, 6, 7,13, 14, 15, 21–26, 28, 35, 57–58, 78, 156–59, 219
pit den, 44, 50, 90–91, 97, 98, 99, 104
Planet Earth III behind the Scenes (film), 33
playing: adults and, 106–7; attacks and, 105–6; behavioral studies and, 106–7; boxing, 46; den studies and, 9, 46, 94, 140; dogs and, 106; novelty and, 108; pups and, 46, 94, *100*, 105–8, 111, 140, 180–81, *plate 2*, *plate 15*; stalking and, 106, 111; toys and, 107–8, 181; wrestling, 105–7, 180; yearlings and, 106
Poland, 177
polar bears (*Ursus maritimus*), xv, 17, 19, 68, 79, 228
Polar Continental Shelf Program (PCSP), 34, 66
Polar Environment Atmospheric Research Lab (PEARL), 68
predation rate, 135, 206, 208
prey, x, xiii, xiv, 2, 6, 11, 12, 24, 26, 27, 28, 29, 30, 39, 41, 54–55, 56, 60, 62, 68, 71, 78, 80, 82, 83, 102, 103, 104, 105, 106, 115, 123, 124–63, 176, 177, 193, 208, 210, 212, 213, 215
Princess Margaret Range, 211
ptarmigan (*Lagopus muta*), 15, 228
pups: aggression and, 111–12, 199; arctic hares and, 99, 104–5; attacks and, 105–6; babysitters for, 15, 26, 51, 150; behavioral studies and, 64–65, 68–72, *plate 18*; birth of, 44, 47, 57, 97, 98, 103, 105, 139; boxing and, 46; breeding females and, 107, 113, 115; breeding males and, 105–7, 112–15, 115; caches and, 116, 118; den studies and, 1–7, 10–15, 20–27, 30, 35, 38–44, 47–49, 53–62, 77, 83, 89–91, 94–118, 130, 140, 144, 149–52, 159–60, 165–68, 171, 175, 178–86, 196–208, 215, *plates 1–3, 6–9, 14–15, 17–19*; Explorer and, 105, 107; food and, 7, 15, 25, 27, 39–42, 56, 60, 65, 94, 96, *100*, 103–5, 108, 111–17, 123, 147, 150, 164–67, 171–74, 183, 208; growth of, 25, 76, 96–98, 104–5, 112, 143, 163, 180–81, 201, 210; headquartering and, 175–83; howling and, *99–100*, 107; hunting and, 112; infanticide and, 196, 205; “licking up” action and, 111, *plate 17*; litter size and, 59, 102, 115, *plate 19*; Mom and, 7, 11–15, 40–44, 47–50, 59, 85, 109–15, 173, 189; mothers and, 5, 7, 15, 39, 43 (*see also* mothers); muskoxen and, 27, 54, 56, 67–68, 72, 81, 86, 94, *100*, 105, 113–16, 123, 130, 143–44, 150,

pups: aggression and (*continued*) 152, 159, 168, 178–79, 183, 187, 208; nursing and, 5, 39–40, 53, 57, 70, 72, 95, 99, 104, 109–12, 138–39, 144, 171–72, 179, 186, 216; observational data on, *91–94, 99–102, 222*; opening of eyes, 90, 98–99, 103–4; playing and, 46, 94, *100*, 105–8, 111, 140, 180–81, *plate 2, plate 15*; regurgitated food and, 26, 39, 49, 99, *100*, 104, 107, 111–15, 123, 159, 166–67, 173, 179, 182; rendezvous sites and, 107; Slidre Fiord and, 77, 81–83; toys and, 107–8, 181; urination and, *99–100*, 108; weaning of, 39, 72, 103, 110–11, 168, 171–73; Whitey and, 105, 111–15; wrestling and, 105–7, 180; yearlings and, 10, 15, 38, 41–43, 54–55, 61, 85, *100*, 105–8, 111–13, 119–20, 130–35, 141, *152*, 170, 172–73, 187–88, 192

purple saxifrage (*Saxifraga oppositifolia*), 19, 136, *146*, 228

Qikiqtani Truth Commission, 19

Quadga Lake, 154

Quebec, 19, 65

Queen Elizabeth Islands, xiii, 5, 18, 62, 129, 136, 145

Queen Maud Gulf, 127

Quttinirpaaq National Park, 18–19

radio collars: arctic hares and, 137; Axel Heiberg Island and, *223, 225*; behavioral studies and, 26–28, 65, 88, 206–7, 215–16; Brutus and, 75–76, 88, 201, 203; Denali National Park and, 12, 39, 53, 75; den studies and, 11–12, 39, 43, 53; GPS, 12, 39, 65, 75, 79–83, 88, 137, 197, 201, 203, 206–7; Inuit people and, 79, 82; muskoxen and, 129; observation data, *223–25*; pioneering use of, 2; range size and, 11–12, 39–40, 83, 137, 207, *226*; Slidre Fiord and, 75–83; territoriality and, 197, 201, 203; tracking technology and, 2, 6, 12, 21, 31, 39, 65, 79, 81, 175

"Rats, Bees, and Brains: The Death of the 'Cognitive Map'" (Goldman), 155

RCMP (Royal Canadian Mounted Police), 80

red deer (*Cervus elaphus*), 177, 228

red foxes (*Vulpes vulpes*), 46, 121

red knot (*Calidris canutus*), 17, 228

regurgitation: behavioral studies and, 165–67, 173; caches and, 117, 123; den studies and, 26, 39, 49, 58; females and, 26, 39, 107, 111–17, 123, 159, 165–66, 179; headquartering and, 179, 182–83; males and, 26, 58, 112, 117, 123, 166–67, 173, 179, 183; mothers and, 39, *100*, 104; pups and, 26, 39, 49, 99, *100*, 104, 107, 111–15, 123, 159, 166–67, 173, 179, 182; yearlings and, 111–13

rendezvous sites: behavioral studies and, 71–73; den studies and, 42–47, 54, 56, 60, 95–97; headquartering and, 181; hunting and, 155, 159; observation data on, *97*; pups and, 107; territoriality and, 197

reproduction: climate change and, 219; den studies and, 62; mating, 63, 65, 103; prey, 103, 143; roles and, 164–67; scent-marking and, 191; survival and, 164; territoriality and, 191–92

research methodology: airplanes, 8–9, 65–66; ATVs, 3 (*see also* ATVs); binoculars, 5, 11–13, 21, 68, 75, 89, 125, 161, 186, 188; blood samples, 11, 31, 83; cameras, 22, 28, 33, 157, 175, 215; darting, 11, 75–78, 81, 83, 88; DNA, 59, 61, 164; drones, 22, 33; effects of, 215; Eureka Weather Station and, *222*; helicopters, 9, 15, 20, 42, 56, 64, 68, 73–74, 77, 80–83, 95,

146; Inuit Hunter Trapper Association (HTA), 79; live-trapping, 2, 11, 31, 190; marking and, 39; mattresses and, 13–15, 21, 24; permits and, 57–58, 79–80, 83, 206; radio collars, xiv, 11–12, 39, 43, 53, 65, 75, 79–83, 88, 129, 137, 197, 201–7, *223–25*; scouting, 5, 7, 18, 74; snowmobiles, 6–9, 12, 18, 24, 61; telescopes, 21–22; terrain, 95, 155; tracking, 6, 12, 21, 31, 39, 70, 74, 79–81, 117
Resolute Bay, xv, 6, 33–34, 62, 65–66
roe deer (*Capreolus capreolus*), 177, 228
roles: breeders, 56, 85, 168–73; den studies and, 38, 56; divisions of labor, 164–74; dominance and, 168–71; food and, 164–67, 171–74; gender, 164; gene promotion and, 164, 171–72; inclusive-fitness theory and, 173–74; kin-selection, 174; leadership, 56, 155, 167–70; reproduction, 63, 65, 103, 164–67; status and, 38
Romulus Lake, 18
rumen, 126

Sawtooth Mountains, 17–18
scats, 5, 59–60, 142, 147
scent: food and, 7, 25, 117, 121–22, 191, 195; ground scratching and, 194–96, 199; hunting and, 156; marking and, 11, 38, 55, 67, 70, 121–22, 156, 186, 191–96, 199; reproduction and, 191; rolling and, 186; territoriality and, 191–96; tracking and, 117; vomeronasal organ (VNO) and, 191–92
Scientific American, 155
scouting, 5, 7, 18, 74
Scruffy, 10
seals (*Phoca* sp.), 17, 228; caches and, 179–80; food and, 51, 145, 148, 172; hunting, 19, 124, 136, 145, 148–49, 207; Inuit people and, 165; skin of, 107; Slidre Fiord and, 148
Shaggy, 159
685M, 174
sled dogs, 216
sleep: arctic hares and, 140–41; behavioral studies and, 86; Brutus and, 78; den studies and, 13, 19, 24–25, 27, 41–42, 49, 55, 57, 96, *100*, 105, 109, 140–41, 196; dogs and, 177; headquartering and, 175–81; hunting and, 140–41; REM, 177; Slidre Fiord and, 78, 80
Slidre Fiord: arctic hares and, 68, 144; breeding females and, 76–77; Brutus and, 74–83; darting and, 75–78, 81, 83; den studies and, 47, 90, 94–95; documentaries and, 205; Eureka Weather Station and, 9, 74, 78; Fosheim Peninsula and, 18–19, 78; foxes and, 77; headquartering and, 178; hunting and, 77–80; muskoxen and, 75, 79, 81–82, 128; National Geographic Society and, 79, 81; pups and, 77, 81–83; radio collars and, 75–83; seals and, 148; sleep and, 78, 80; terrain of, 18; tracking and, 74, 79–81; traps and, 79; tundra and, 77–78; Whitey and, 47
Slidre River, 44, 74, 95
snarling, 55, 173
snowmobiles, 6–9, 12, 18, 24, 61
Snow Wolf Family and Me (BBC), 32
snowy owl (*Bubo scandiacus*), 17, 228
Soviet Union, 20
spotting scopes, 2, 12, 68, 74–75, 89, 95
stalking: food and, 51–52; hunting and, 139, 157, 160–62; play, 106, 111
"standing over" posture, 111
starvation, 60–62, 116, 210–11
status: alpha females, 63, 173; alpha males, 63, 158–59, 173, 176, 181, 183, 199; "Alpha Status, Dominance, and

status: alpha females (*continued*)
Division of Labor in Wolf Packs," 159, 174; behavioral studies and, 174; den studies and, 38, 46–47, 63; hierarchy, 63; hunting and, 159; leadership, 56, 155, 167–70; marking and, 191; Mom and, 43–47; roles and, 38; submission and, 38, 46, 48–49, 159, 170–73, 180; territoriality and, 191–92; trespassers, 11, 192, 196, 205, 207
submission: "licking up" action and, 46, 49, 111–12, 159, 165–67, 170, 173, 180, 182–83, *plate 17*; status and, 38, 46, 48–49, 159, 170–73, 180
Superior National Forest, xiv, 36, 150–54
SuperPup, 44, 73
survival: arctic hares and, 67–68, 216; behavior and, 164; competition for, 147; food and, 213 (*see also* food); genetic, 164, 172; inclusive-fitness theory and, 173–74; litter sizes and, 59, 102, 115, *plate 19*; mortality and, 2, 69–70, 82, 96, 208, 211; muskoxen and, 67–68; predator search time and, 139; reproduction and, 164; tracking, 82
Sweden, 135

tables, 45, 69, 85, 91–94, 97, 99–100, 101, 102, 113, 114, 131, 146, 152, 211, 222, 223, 224, 225, 226
tail wagging, 151, 185
tapirs (*Tapirus* sp.), 17, 228
telescopes, 21–22
temperatures, 6, 16–17, 211–13
terrain traps, 155
territoriality: aggression and, 194–95, 199; alpha males and, 199; attacks and, 194; Axel Heiberg Island and, 201, 203, 205; boundaries and, 193–96; breeding females and, 192, 194, 204; breeding males and, 192; Brutus and, 201–7; Denali National Park and, 196; dominance and, 191–92; Eureka Weather Station and, 194, 197–201, 204; food and, 191, 195, 197; Fosheim Peninsula and, 201–5; ground scratching and, 194–96, 199; howling and, 184, 194–98; hunting and, 193; infanticide and, 196, 205; marking and, 11, 67, 70, 121–23, 156, 179, 191–99; Mom and, 196–97; mothers and, 192; muskoxen and, 194–95, 199, 205; radio collars and, 197, 201, 203; range size and, 11–12, 39–40, 137, 207, *226*; rendezvous sites and, 197; reproduction and, 191–92; scent and, 191–96; status and, 191–92; trespassers, 11, 40, 58, 168, 192, 196–201, 205, 207; vomeronasal organ (VNO) and, 191–92; Whitey and, 195–201; yearlings and, 192; Yellowstone and, 193, 196, 201, 205
testicle, 182
testing, by wolves of prey, 134, 155, 160, 178
tests, of wolves, 12, 13, 15, 25, 31, 39, 57, 112, 114, 129, 141, 165, 171–73, 179, 185, 188–90, 194, 198, 215
theory of mind, 40–41, 153–55, 160–62, 187
thermal oasis, 116
Thule, Greenland, xiii, 19, 34, 79
Thule people, 19
tiger (*Panthera tigris*), 228
topography, x, 5, 6, 13, 16, 17, 18, 40, 68, 71, 89, 92, 132, 136, 137, 139, 146, 178, 229, 241
toys, 107–8, 181
tracking: behavioral studies and, 70; caches and, 117; den studies and, 12, 21, 31, 39; radio collars, 2, 6, 12, 21, 31, 39, 65, 79, 81, 175; scats and, 5, 59–60, 142, 147; Slidre Fiord and, 74, 79–81; snow, 21; USFWS and, 6; wind and, 189

traps: behavioral studies and, 168; camera, 215; den studies and, 2, 11, 21, 31, 95; foot, 11; foxes and, 154; Inuit Hunter Trapper Association (HTA), 79; Iviq Hunters and Trappers Association, 207, 219; live, 2, 11, 31, 190; Slidre Fiord and, 79; terrain, 155
travel speed, 151, *152*
trespassing: Brutus and, 205, 207; human, 58; protecting pups and, 40; responses to, 11, 168, 198–201, 205; scent-marking and, 192; territoriality, 11, 40, 58, 168, 192, 196–201, 205, 207
tropical habitats, x, 17
trust, 24–25, 35, 167
tundra: arctic hares and, 136–37, 139; behavioral studies and, 86–88; climate change and, 213; den studies and, 13, 18, 27–30, 36, 40, 95, 97, *plate 2*; Ellesmere Island and, 3, 5; headquartering and, 180; howling and, 189; hunting and, 86, 151; muskoxen and, 130; Slidre Fiord and, 77–78
Turner Productions, 32
Twin Otter, 66, 68

urination: caches and, 121–22; den studies and, 35, 46; flexed-leg (FL), 121–22, 192–95; hunting and, 156–58; marking by, 46, 121–22, 156, 192–94; pups and, *99–100*, 108; raised-leg (RL), 46, 121–22, 156, 173, 181, 192–94; squat, 122, 156, 158, 165, 194; standing, 108, 122, 182, 189, 193
US Fish and Wildlife Service (USFWS), 6, 35, 103
US National Park Service, 15
Utah State University, xiv

vegetation: arctic hares and, 136, 138; climate change and, 213; den studies and, 15, 17–19, 57, 96; Ellesmere Island and, 17–18; as food, 17, 129, 155, 210, 213; muskoxen and, 129; as protection, 96; terrain and, 157, 178
videos, 32, 35, 87, 140, 147, 169, 199
vomeronasal organ (VNO), 191–92

walrus (*Odobenus rosmarus*), 17, 228
weaning, 39, 72, 103, 110–11, 168, 171–73
weather, 4, 6, 13, 14, 15, 16, 23, 36, 66, 82, 89, 90, 91, 102, 135, 210, 211, 212, 223, 225, 226, 235, 237
weather station, 4, 6, 7, 9, 10, 18, 20, 21, 24, 33, 34, 40, 43, 47, 53, 54, 58, 60, 61, 62, 66, 68, 70, 72, 73, 74, 78, 88, 95, 96, 107, 124, 138, 146, 147, 166, 178, 194, 197, 198, 199, 201, 204, 212, 216, 219, 222. *See also* Eureka Weather Station
White Falcon, White Wolf (PBS), 32
White Wolf (National Geographic Explorer), 32–33, 87, 147, 152, 159, 199
"White Wolves: Ghosts of the Arctic" (CBC), 33
Whitey: arctic hares and, 139; behavioral studies and, 46, 85–86, 165–73; caches and, 118, 122–23; den studies and, 38, 42–51, 54–61, 90–94; headquartering and, 177–83; howling and, 184–90; Mom and, 42–50, 59, 85, 115, 122, 173, 184, 196–97; pups and, 105, 111–15; Slidre Fiord and, 47; territoriality and, 195–201
wild boar (*Sus scrofa*), 177, 228
Wild Wolves We Have Known (Thiel, Thiel, and Strozewski), 84–88
wind: constant, 16; dressing for, 6–7; frigid, 175; shearing, 83; strong, 36, 186; tracking and, 189; weathering from, 23
wolf (*Canis lupus*), 228
Wolf, The: The Ecology and Behavior of an Endangered Species (Mech), xiv, 63
Wolf A, 160–61

Wolf B, 160–61
Wolf Science Center, 41
Wolf Specialist Group, 36, 44, 215
wolverine (*Gulo gulo*), xv, 228
Wolves: Behavior, Ecology and Conservation (Mech and Boitani), 215
wolves, fear of, 78, 87
wolves, lack of fear, 6–9, 46–47, 87
Wolves of Mount McKinley, The (Murie), 1
Wolves on the Hunt (Mech, Smith, and MacNulty), 129, 134
World Wildlife Fund, 212–13
wrestling, 105–7, 180

yearlings: arctic hares and, 105, 141; behavioral studies and, 85, 170–73; caches and, 119–20; den studies and, 10, 15, 38, 41–43, 54–55, 61; howling and, 187–88; hunting and, 152; lemmings and, 108; muskoxen and, 41, 130, *131*, 133–35; observation data on, *100*; playing and, 106; regurgitation and, 111–13; territoriality and, 192; travel speeds of, *152*
Yellowknife, 34, 65, 70
Yellowstone National Park: alphas and, 63; behavioral studies and, 169–170, 173, 208, 214; climate change and, 205; elk and, 155; headquartering and, 175; howling and, 185; McIntyre and, 40, 54, 117, 169–70, 185, 193; reintroduction of wolves in, 12, 22, 36, 67, 201, 214; territoriality and, 193, 196, 201, 205; 06 Female, 40
Yellowstone Wolf Project, ix, xiv
YouTube, 32–33, 51, 76, 140, 159

zigzagging, 117–18, 137, 140, 198

Printed and bound by CPI Group (UK) Ltd, Croydon, CR0 4YY

16/04/2025

14658566-0002